Student Edition

Eureka Math

Grade 1
Modules 3 & 4

Special thanks go to the Gordon A. Cain Center and to the Department of Mathematics at Louisiana State University for their support in the development of *Eureka Math*.

For a free *Eureka Math* Teacher Resource Pack, Parent Tip Sheets, and more please visit www.Eureka.tools

Published by the non-profit Great Minds

Printed in the U.S.A.
This book may be purchased from the publisher at eureka-math.org
10 9

ISBN 978-1-63255-290-7

Name Liachatter Date ____________

Write the words **longer than** or **shorter than** to make the sentences true.

1.

Abby

Spot

Abby is shorter than Spot.

2.

A B

B is shorter than A.

3.

The American flag hat

is ______________________

the chef hat.

4.

The darker bat's wingspan

is ______________________

the lighter bat's wingspan.

5.

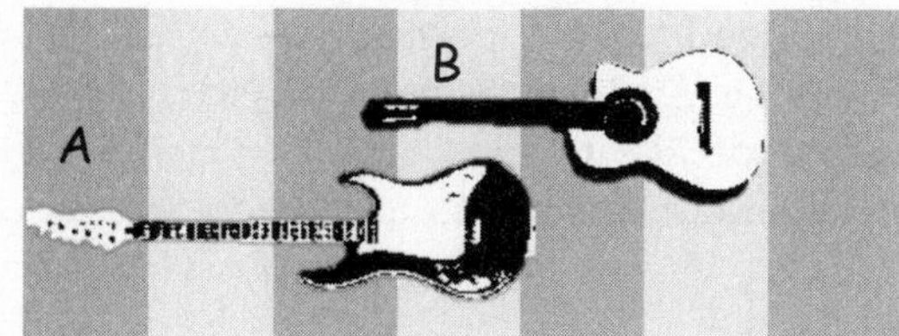

Guitar B is

____

Guitar A.

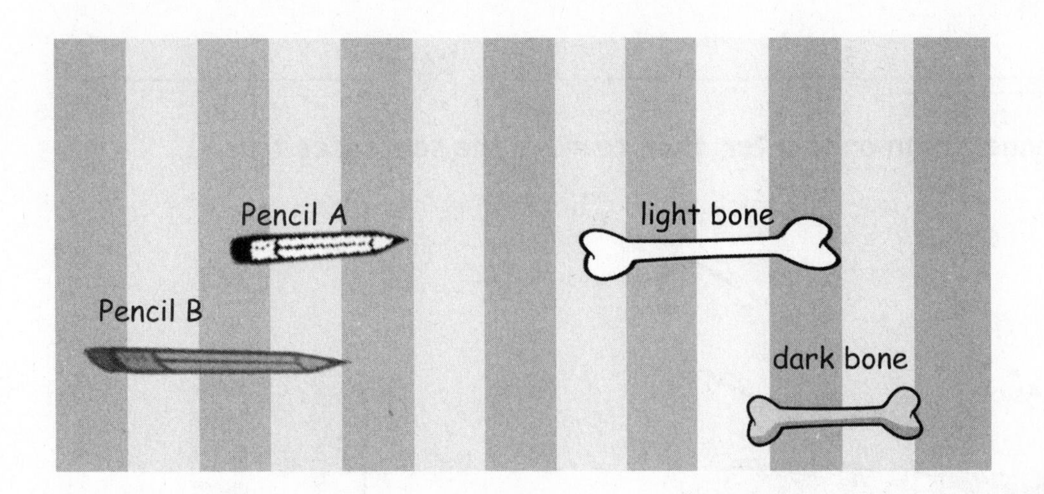

6. Pencil B is ________________________ Pencil A.

7. The dark bone is ________________________ the light bone.

8. Circle true or false.
   The light bone is shorter than Pencil A. **True** or **False**

---

9. Find 3 school supplies. Draw them here in order from **shortest** to **longest**.
   Label each school supply.

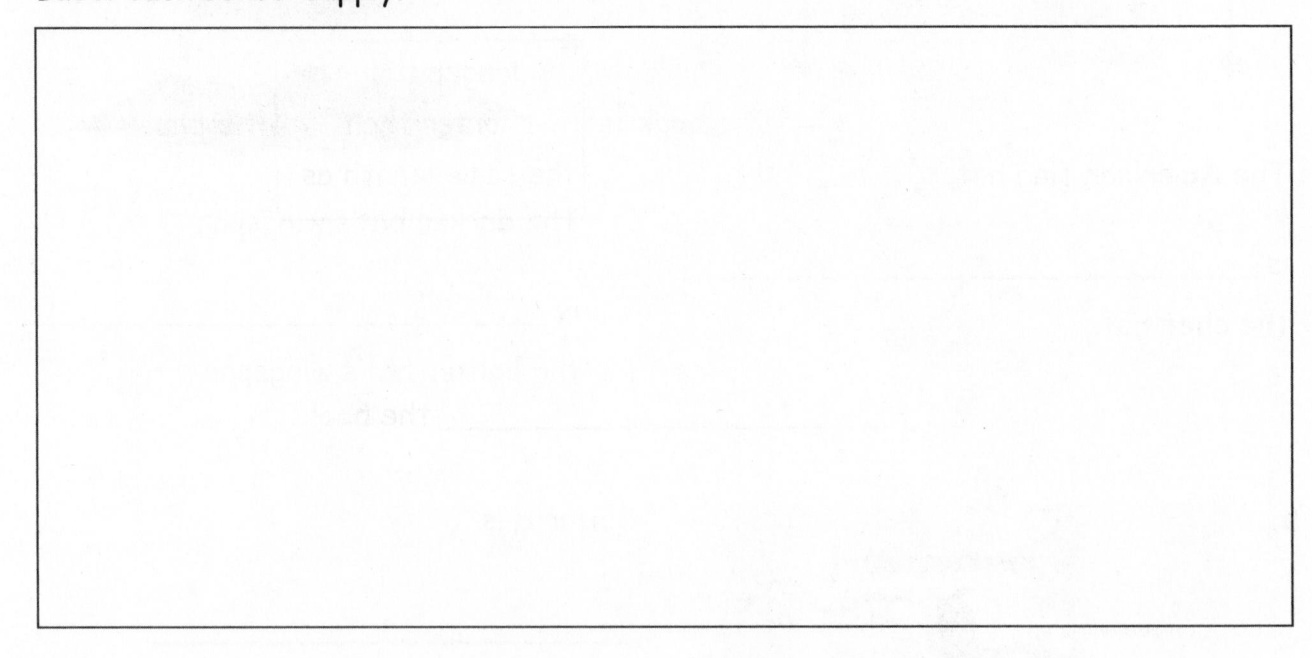

Name ______________________________ Date ______________

1. Use the paper strip provided by your teacher to measure each **picture**. Circle the words you need to make the sentence true. Then, fill in the blank.

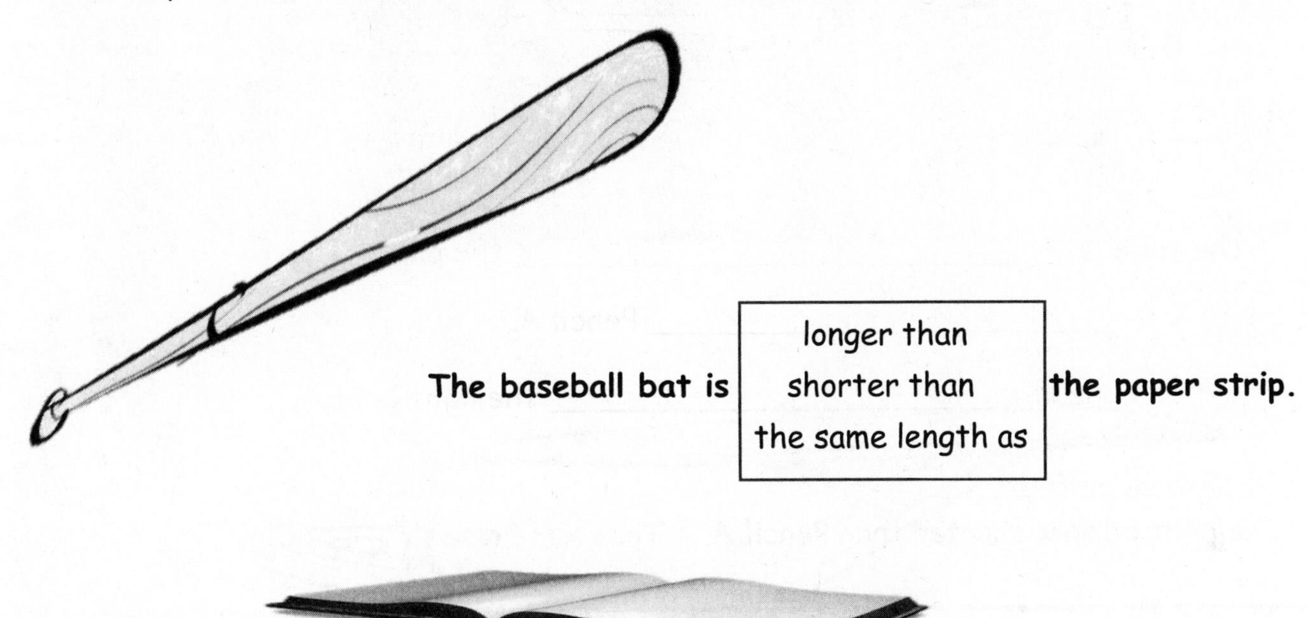

The baseball bat is | longer than / shorter than / the same length as | **the paper strip.**

The book is | longer than / shorter than / the same length as | **the paper strip.**

The **baseball bat** is ______________________________ the **book.**

2. Complete the sentences with **longer than**, **shorter than**, or **the same length as** to make the sentences true.

a.

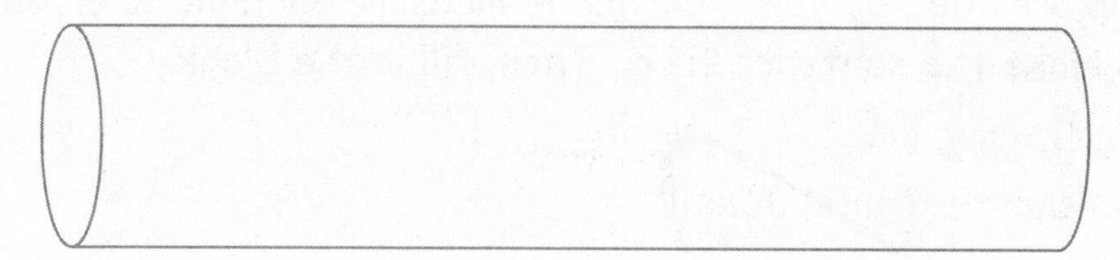

The **tube** is ______________________________ the **cup**.

b.

The **iron** is ______________________________ the **ironing board**.

Use the measurements from Problems 1 and 2. Circle the word that makes the sentences true.

3. The baseball bat is (**longer/shorter**) than the cup.

4. The cup is (**longer/shorter**) than the ironing board.

5. The ironing board is (**longer/shorter**) than the book.

6. Order these objects from shortest to longest:

cup, tube, and paper strip

____________________ ____________________ ____________________

Draw a picture to help you complete the measurement statements. Circle the words that make each statement true.

7. Sammy is taller than Dion.
   Janell is taller than Sammy.
   Dion is (**taller than/shorter than**) Janell.

8. Laura's necklace is longer than Mihal's necklace.
   Laura's necklace is shorter than Sarai's necklace.
   Sarai's necklace is (**longer than/shorter than**) Mihal's necklace.

Name ______________________________ Date ______________

Use the paper strip provided by your teacher to measure each **picture**. Circle the words you need to make the sentence true. Then, fill in the blank.

1.

The sundae is [longer than / shorter than / the same length as] **the paper strip.**

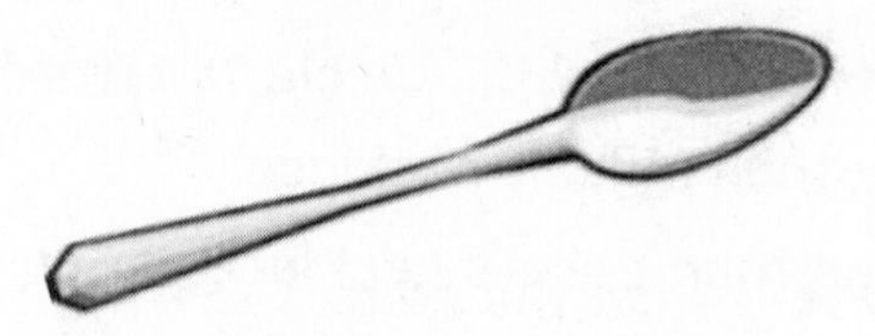

The spoon is [longer than / shorter than / the same length as] **the paper strip.**

The **spoon** is ______________________________ the **sundae**.

2.

The **balloon** is ______________________________ the **cake**.

3.

The **ball** is shorter than the paper strip.

So, the **shoe** is ______________________ the **ball**.

Use the measurements from Problems 1–3. Circle the word that makes the sentences true.

4. The spoon is (**longer/shorter**) than the cake.

5. The balloon is (**longer/shorter**) than the sundae.

6. The shoe is (**longer/shorter**) than the balloon.

7. Order these objects from shortest to longest:

   cake, spoon, and paper strip

   ______________ ______________ ______________

Draw a picture to help you complete the measurement statements. Circle the word that makes each statement true.

8. Marni's hair is shorter than Wesley's hair.
   Marni's hair is longer than Bita's hair.
   Bita's hair is (**longer/shorter**) than Wesley's hair

9. Elliott is shorter than Brady.
   Sinclair is shorter than Elliott.
   Brady is (**taller/shorter**) than Sinclair.

If ______________ is longer than
(classroom object)
my foot and

______________ is shorter than my
(classroom object)
foot, then

______________ is longer than
(classroom object)

______________.
(classroom object)

My foot is about the same length as ______________.
(classroom object)

indirect comparison statements

This page intentionally left blank

Name ______________________________ Date ______________

1. In a playroom, Lu Lu cut a piece of string that measured the distance from the doll house to the park. She took the same string and tried to measure the distance between the park and the store, but she ran out of string!

   Which is the longer path? Circle your answer.

the doll house to the park

the park to the store

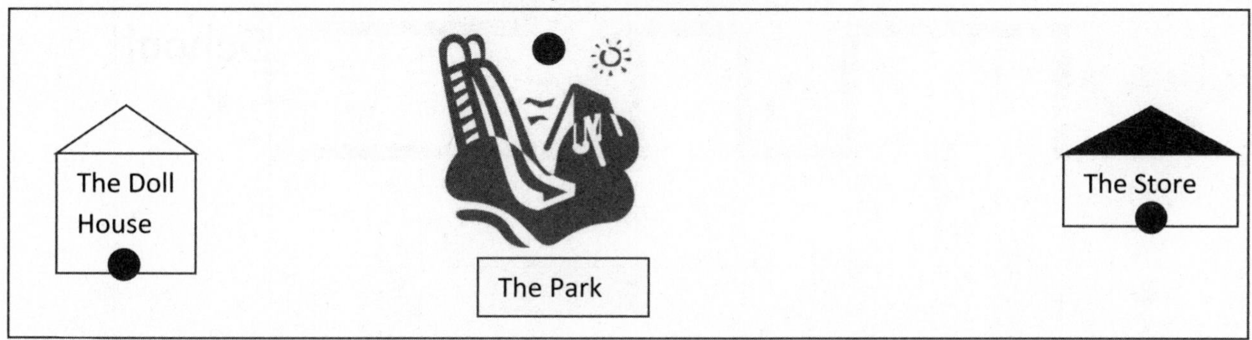

Use the picture to answer the questions about the rectangles.

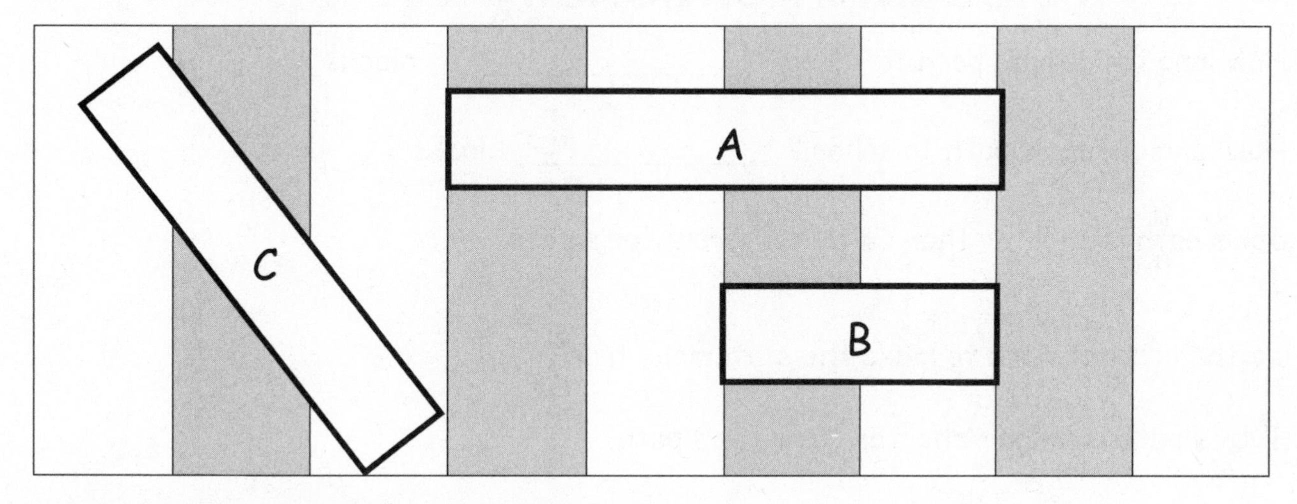

2. Which is the shortest rectangle? ________________

3. If Rectangle A is longer than Rectangle C, the longest rectangle is ___________.

4. Order the rectangles from shortest to longest:

______________ ______________ ______________

Use the picture to answer the questions about the students' paths to school.

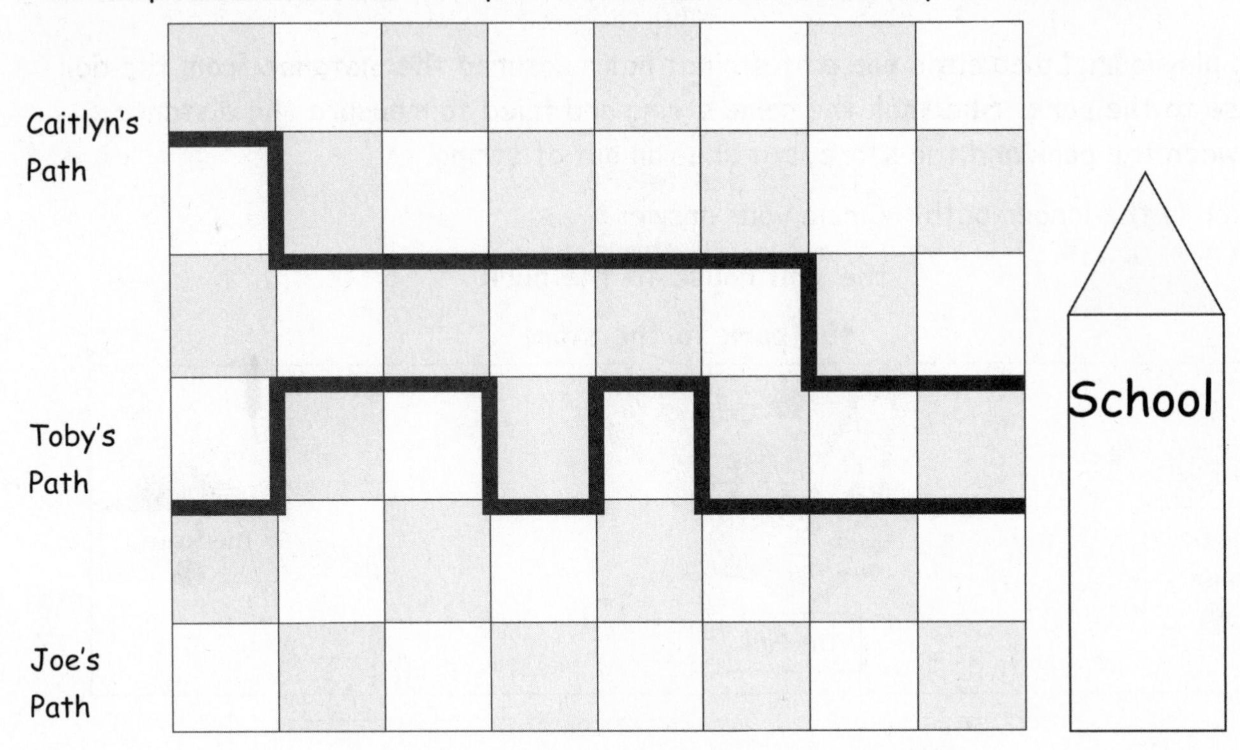

5. How long is Caitlyn's path to school? ________________ blocks

6. How long is Toby's path to school? _____________ blocks

7. Joe's path is shorter than Caitlyn's. Draw Joe's path.

Circle the correct word to make the statement true.

8. Toby's path is **longer/shorter** than Joe's path.

9. Who took the shortest path to school? ______________________

10. Order the paths from shortest to longest.

________________ ________________ ________________

Name ______________________________ Date ______________

1. The string that measures the path from the garden to the tree is longer than the path between the tree and the flowers. Circle the shorter path.

the garden to the tree

the tree to the flowers

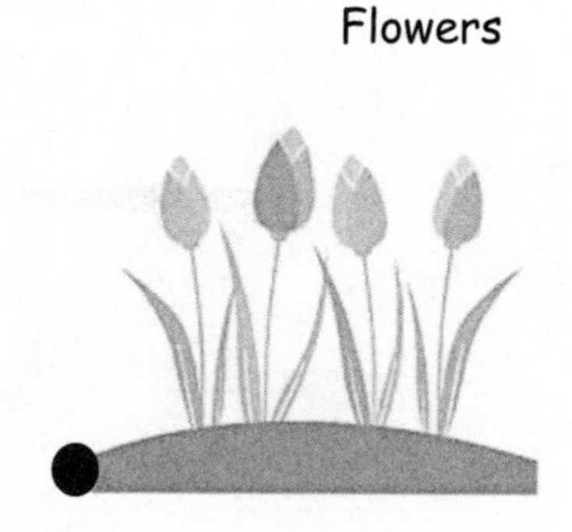

Use the picture to answer the questions about the rectangles.

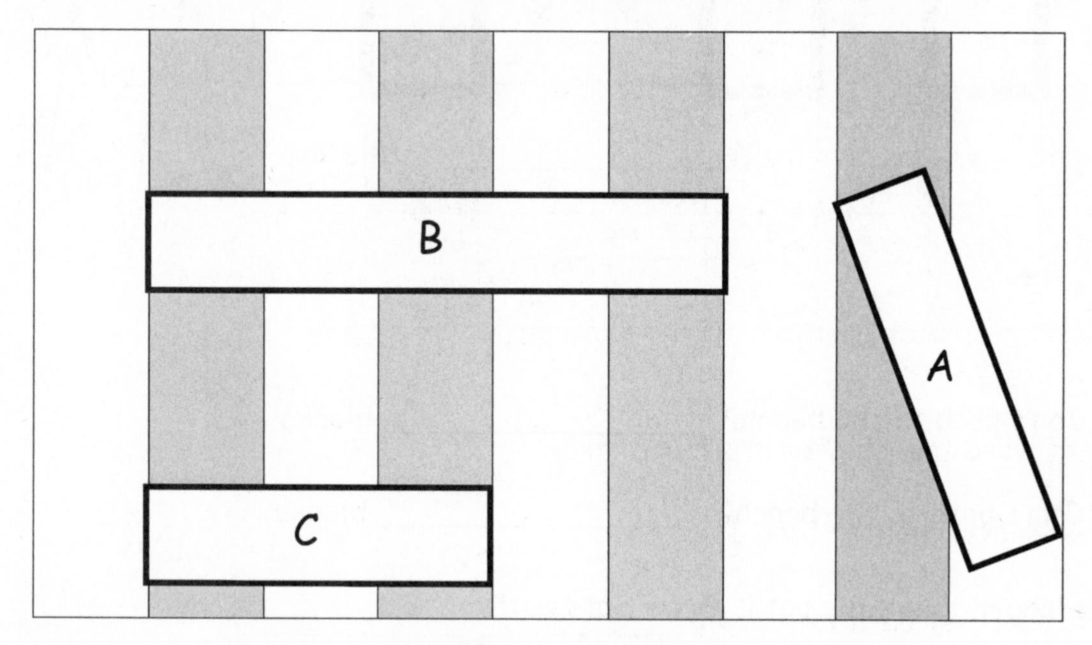

2. Which is the longest rectangle? ________________

3. If Rectangle A is longer than Rectangle C, the shortest rectangle is

________________.

4. Order the rectangles from shortest to longest.

________________ ________________ ________________

Use the picture to answer the questions about the children's paths to the beach.

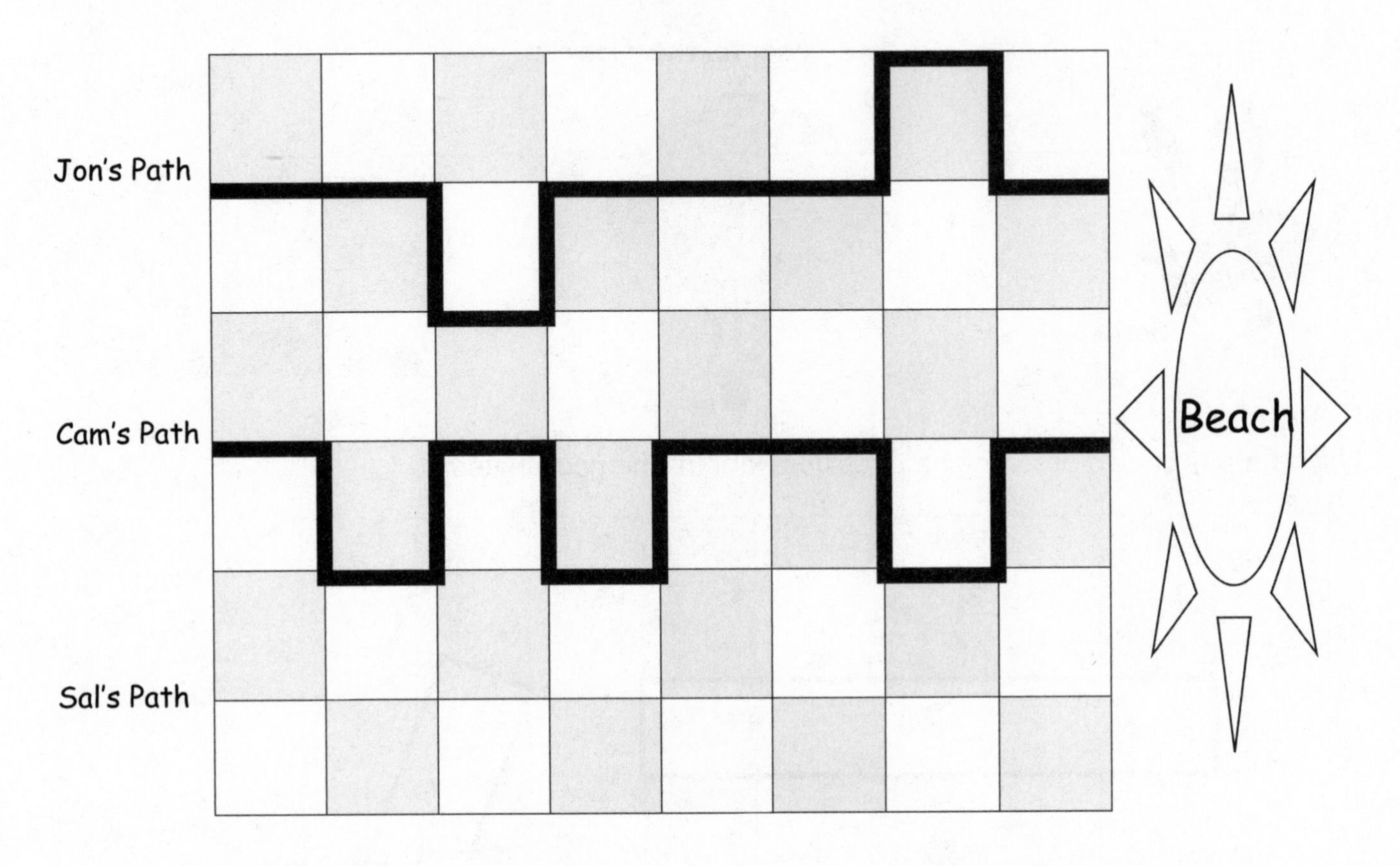

5. How long is Jon's path to the beach? ________________ blocks

6. How long is Cam's path to the beach? ________________ blocks

7. Jon's path is longer than Sal's path. Draw Sal's path.

Circle the correct word to make the statement true.

8. Cam's path is **longer/shorter** than Sal's path.

9. Who took the shortest path to the beach? ______________

10. Order the paths from shortest to longest.

______________ ______________ ______________

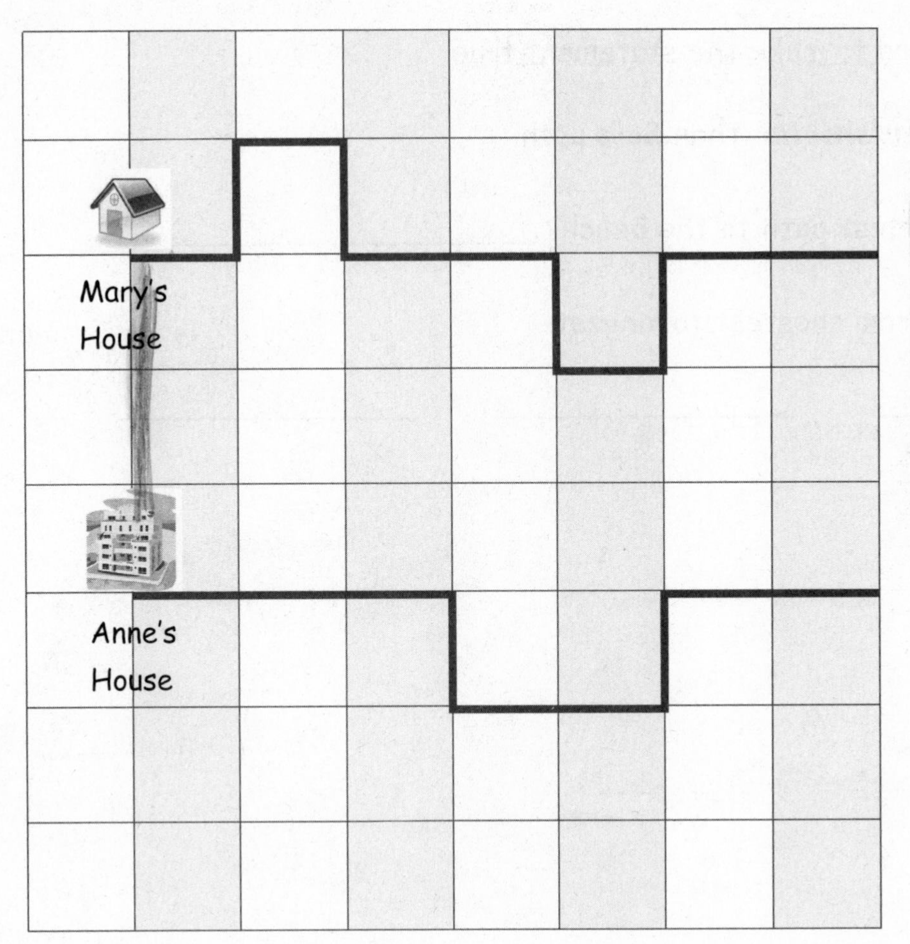

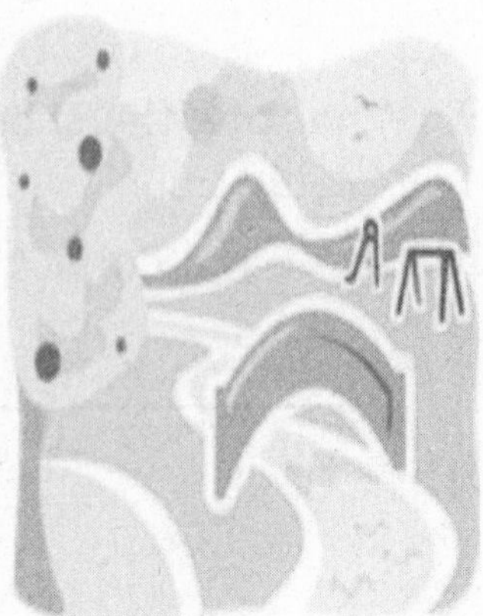

city blocks grid

Name ______________________________ Date ______________

Measure the length of each picture with your cubes. Complete the statements below.

1. The pencil is ______ centimeter cubes long.

2. The pan is ______ centimeter cubes long.

3. The shoe is ______ centimeter cubes long.

4. The bottle is ______ centimeter cubes long.

5. The paintbrush is ______ centimeter cubes long.

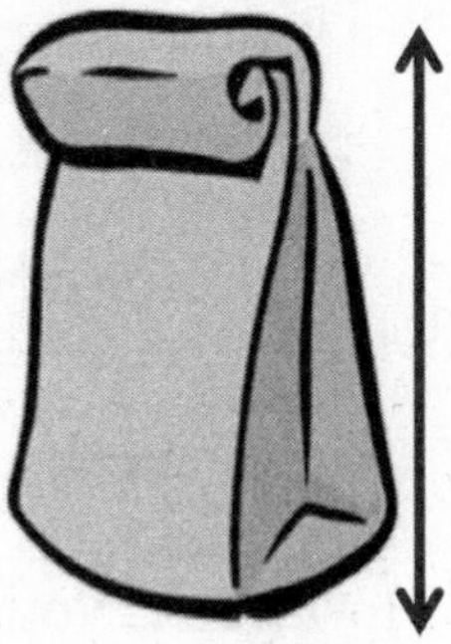

6. The bag is ______ centimeter cubes long.

7. The ant is ______ centimeter cubes long.

8. The cupcake is ______ centimeter cubes long.

9.

The cow sticker is _______ centimeter cubes long.

10.

The vase is _______ centimeter cubes long.

11. Circle the picture that shows the correct way to measure.

A

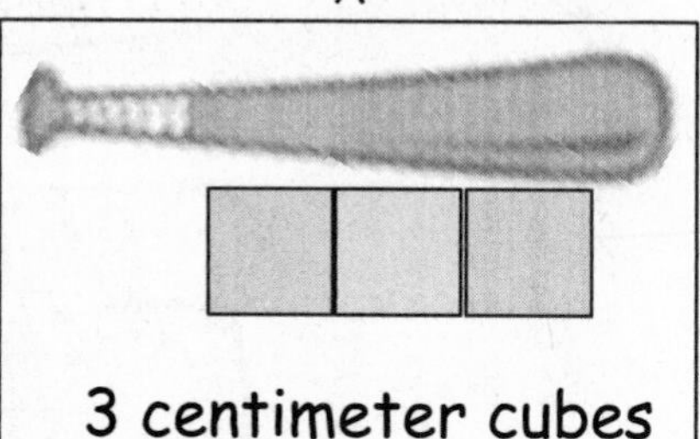

3 centimeter cubes

B

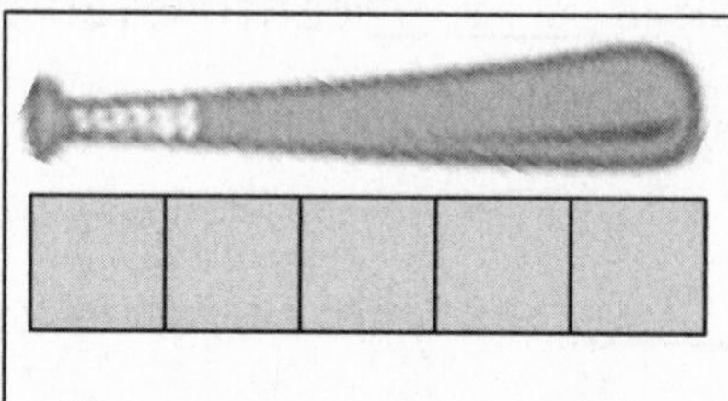

5 centimeter cubes

12. How would you fix the picture that shows an incorrect measurement?

____

____

Name ______________________ Date ____________

Measure the length of each picture with your cubes. Complete the statements below.

1. The lollipop is ______ centimeter cubes long.

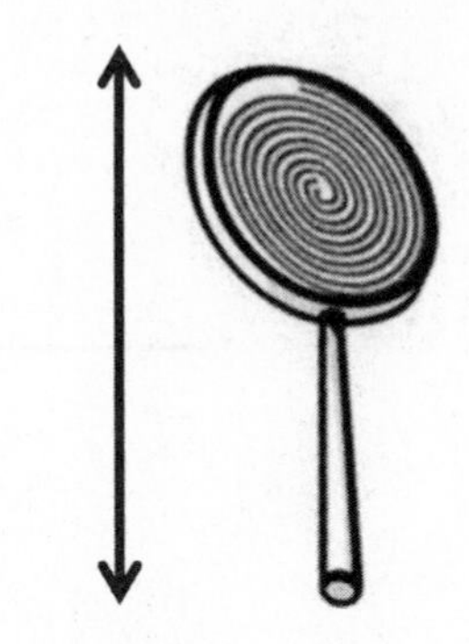

2. The stamp is ______ centimeter cubes long.

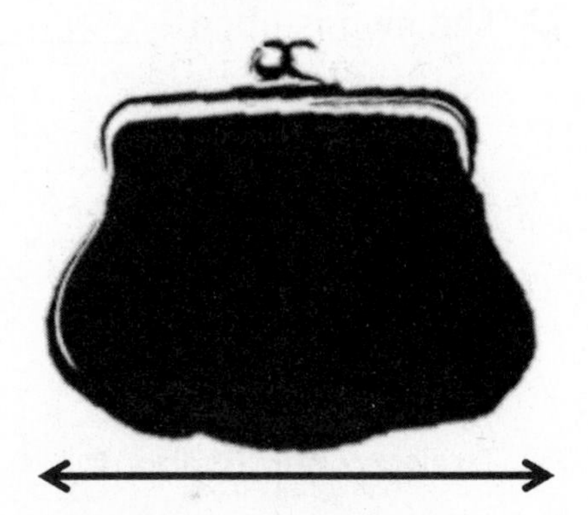

3. The purse is ______ centimeter cubes long.

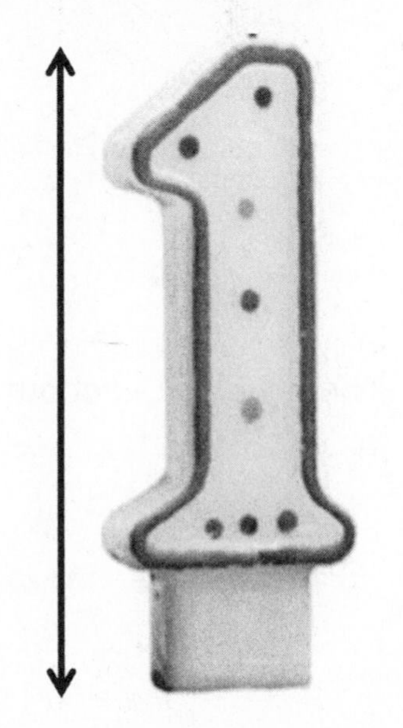

4. The candle is ______ centimeter cubes long.

5. The bow is ______ centimeter cubes long.

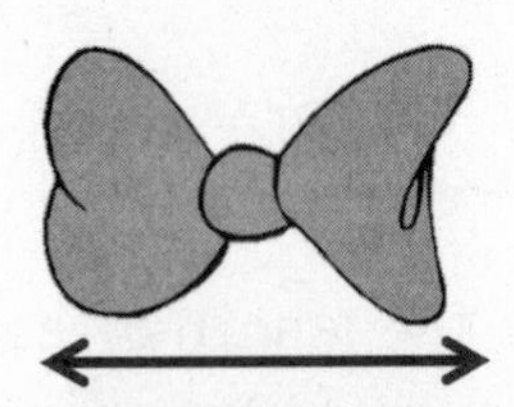

6. The cookie is ______ centimeter cubes long.

7. The mug is about ______ centimeter cubes long.

8. The ketchup is about ______ centimeter cubes long.

9. The envelope is about ______ centimeter cubes long.

10. Circle the picture that shows the correct way to measure.

A

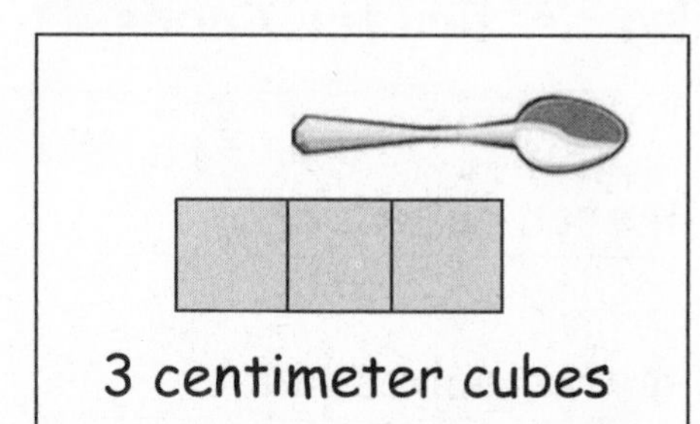

B

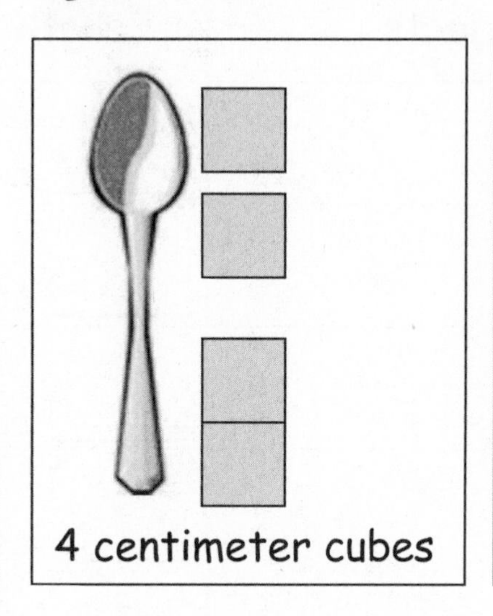

C

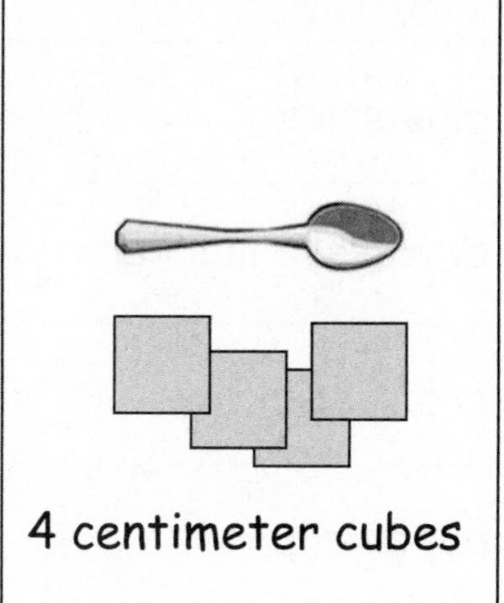

D

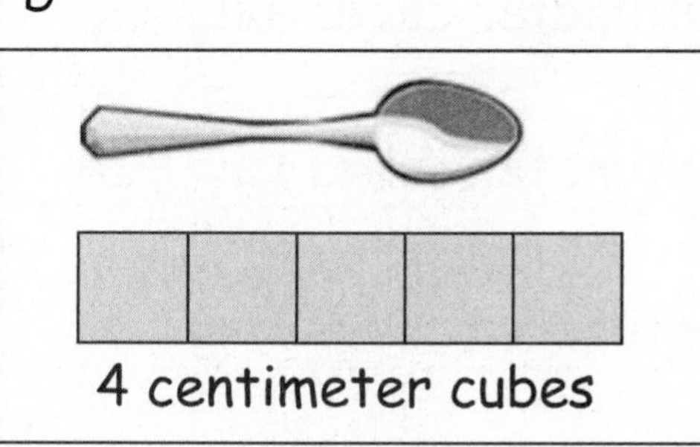

11. Explain what is wrong with the measurements for the pictures you did NOT circle.

____

____

____

Name ______________________________ Date ______________

Classroom Objects	Length Using Centimeter Cubes
glue stick	_____ centimeter cubes long
dry erase marker	_____ centimeter cubes long
craft stick	_____ centimeter cubes long
paper clip	_____ centimeter cubes long
	_____ centimeter cubes long
	_____ centimeter cubes long
	_____ centimeter cubes long

measurement recording sheet

Name ______________________________ Date ______________

1. Circle the object(s) that are measured correctly.

a.

3 centimeters long

b.

5 centimeters long

c.

4 centimeters long

2. Measure the paper clip in 1(b) with your cubes. Then, check the cubes with your centimeter ruler.

The paper clip is __________ *centimeter cubes* long.

The paper clip is __________ *centimeters* long.

Be ready to explain why these are the same or different during the Debrief!

3. Use centimeter cubes to measure the length of each picture from left to right. Complete the statement about the length of each picture in centimeters.

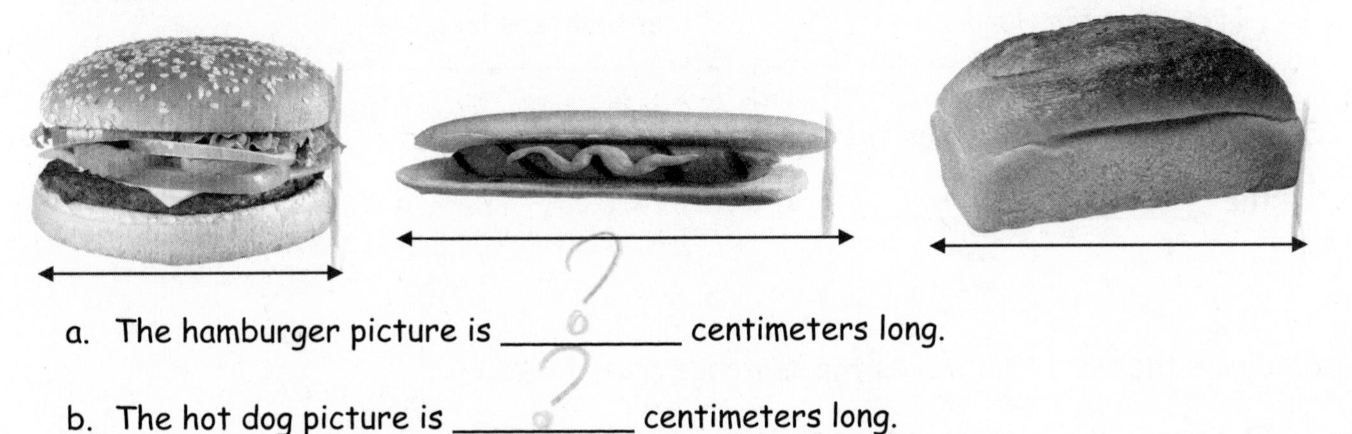

a. The hamburger picture is __________ centimeters long.

b. The hot dog picture is __________ centimeters long.

c. The bread picture is __________ centimeters long.

4. Use centimeter cubes to measure the objects below. Fill in the length of each object.

a.
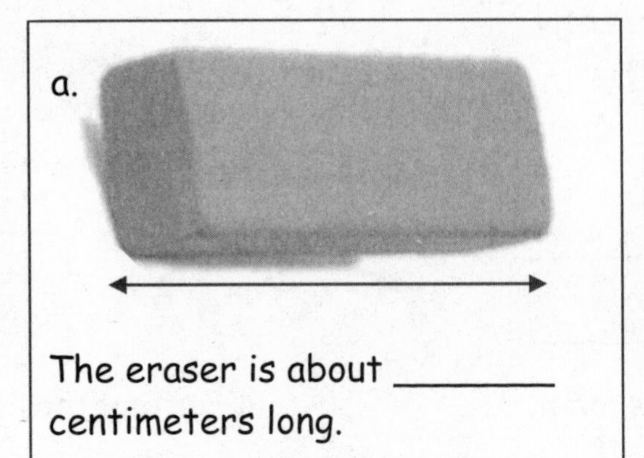
The eraser is about _______ centimeters long.

b.
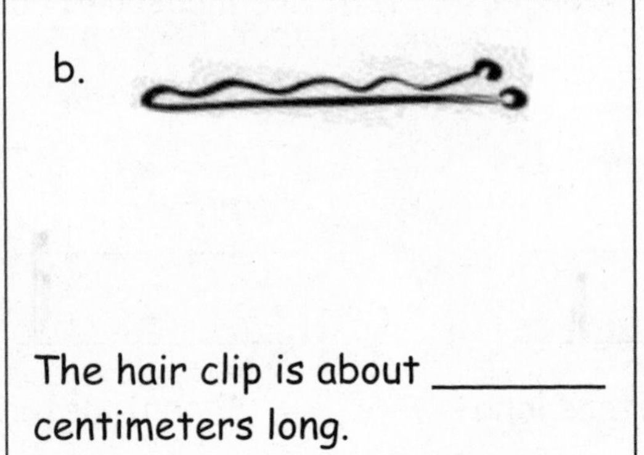
The hair clip is about _______ centimeters long.

c.
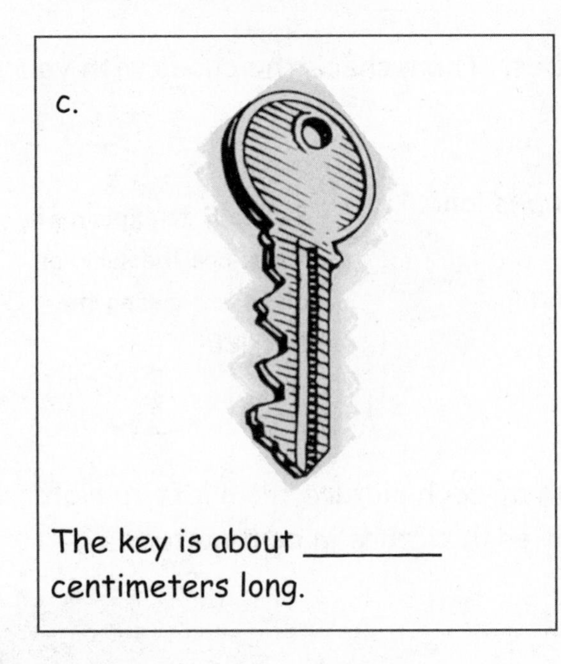
The key is about _______ centimeters long.

d.

The marker is about _______ centimeters long.

5. The eraser is longer than the ____________________, but it is shorter than the ____________________.

6. Circle the word that makes the sentence true.

If a paper clip is shorter than the key, then the marker is **longer/shorter** than the paper clip.

Name ______________________________ Date ______________

1. Justin collects stickers. Use centimeter cubes to measure Justin's stickers. Complete the sentences about Justin's stickers.

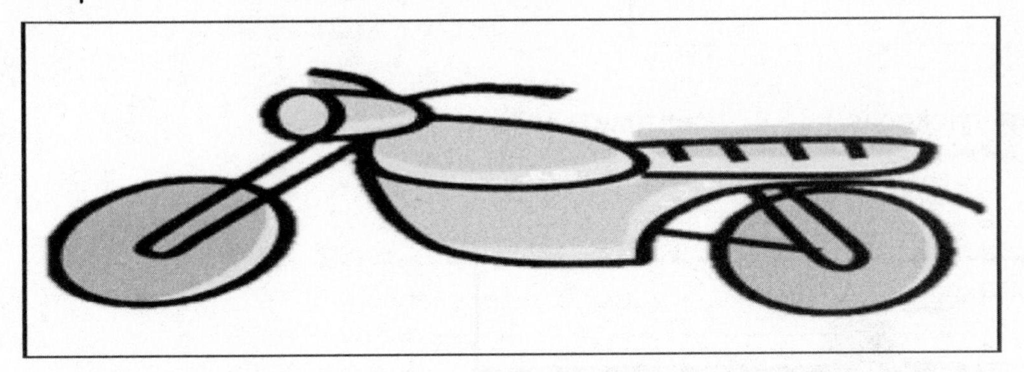

a. The motorcycle sticker is ______ centimeters long.

b. The car sticker is ______ centimeters long.

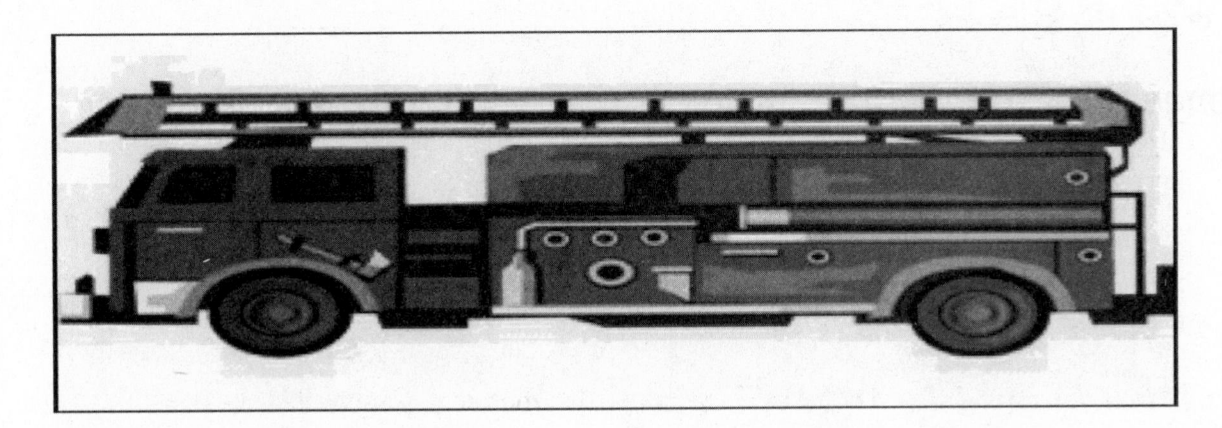

c. The fire truck sticker is ______ centimeters long.

d. The rowboat sticker is ______ centimeters long.

e. The airplane sticker is ______ centimeters long.

2. Use the stickers' measurements to order the stickers of the **fire truck**, the **rowboat**, and the **airplane** from longest to shortest. You can use drawings or names to order the stickers.

Longest ⟶ Shortest

3. Fill in the blanks to make the statements true. (There may be more than one correct answer.)

   a. The airplane sticker is longer than the _______________ sticker.

   b. The rowboat sticker is longer than the ________________ sticker and shorter than the _______________ sticker.

   c. The motorcycle sticker is shorter than the _____________ sticker and longer than the _____________ sticker.

   d. If Justin gets a new sticker that is longer than the rowboat, it will also be longer than which of his other stickers? _________________

This page intentionally left blank

Name ______________________________ Date ______________

1. Order the bugs from longest to shortest by writing the bug names on the lines. Use centimeter cubes to check your answer. Write the length of each bug in the space to the right of the pictures.

The bugs from longest to shortest are

_______________ _______________ _______________

Fly

____ centimeters

Caterpillar

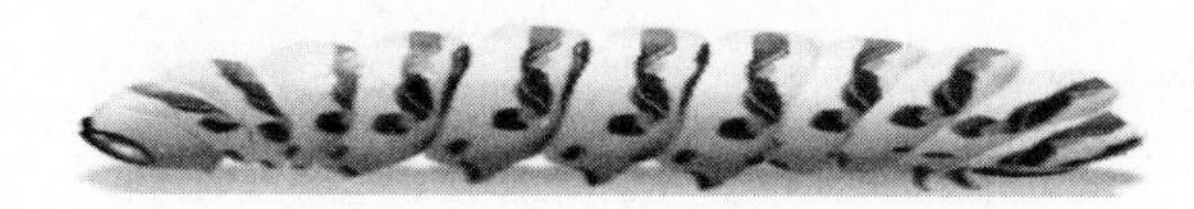

____ centimeters

Bee

____ centimeters

2. Order the objects below from shortest to longest using the numbers 1, 2, and 3. Use your centimeter cubes to check your answers, and then complete the sentences for problems d, e, f, and g.

   a. The noise maker: ________

   b. The balloon: ________

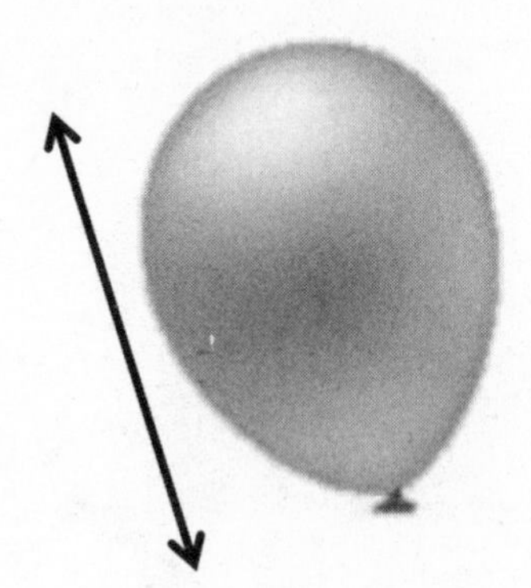

   c. The present: ________

   d. The present is about ________ centimeters long.

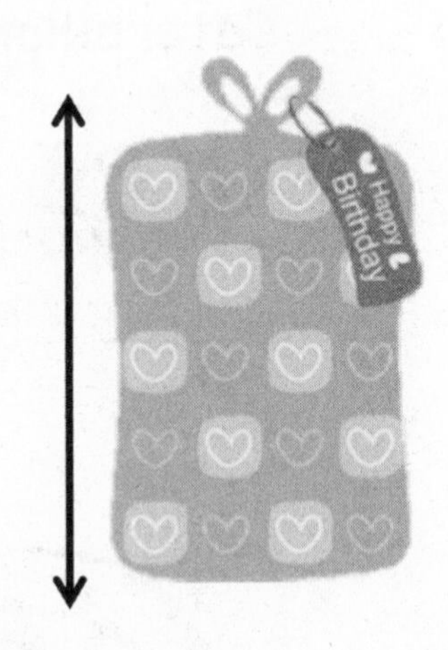

   e. The noise maker is about ________ centimeters long.

   f. The balloon is about ________ centimeters long.

   g. The noise maker is about ________ centimeters longer than the present.

Use your centimeter cubes to model each length, and answer the question. Write a statement for your answer.

3. Peter's toy T. rex is 11 centimeters tall, and his toy Velociraptor is 6 centimeters tall. How much taller is the T. rex than the Velociraptor?

4. Miguel's pencil rolled 17 centimeters, and Sonya's pencil rolled 9 centimeters. How much less did Sonya's pencil roll than Miguel's?

5. Tania makes a cube tower that is 3 centimeters taller than Vince's tower. If Vince's tower is 9 centimeters tall, how tall is Tania's tower?

Name ____________________ Date __________

1. Natasha's teacher wants her to put the fish in order from longest to shortest. Measure each fish with the centimeter cubes that your teacher gave you.

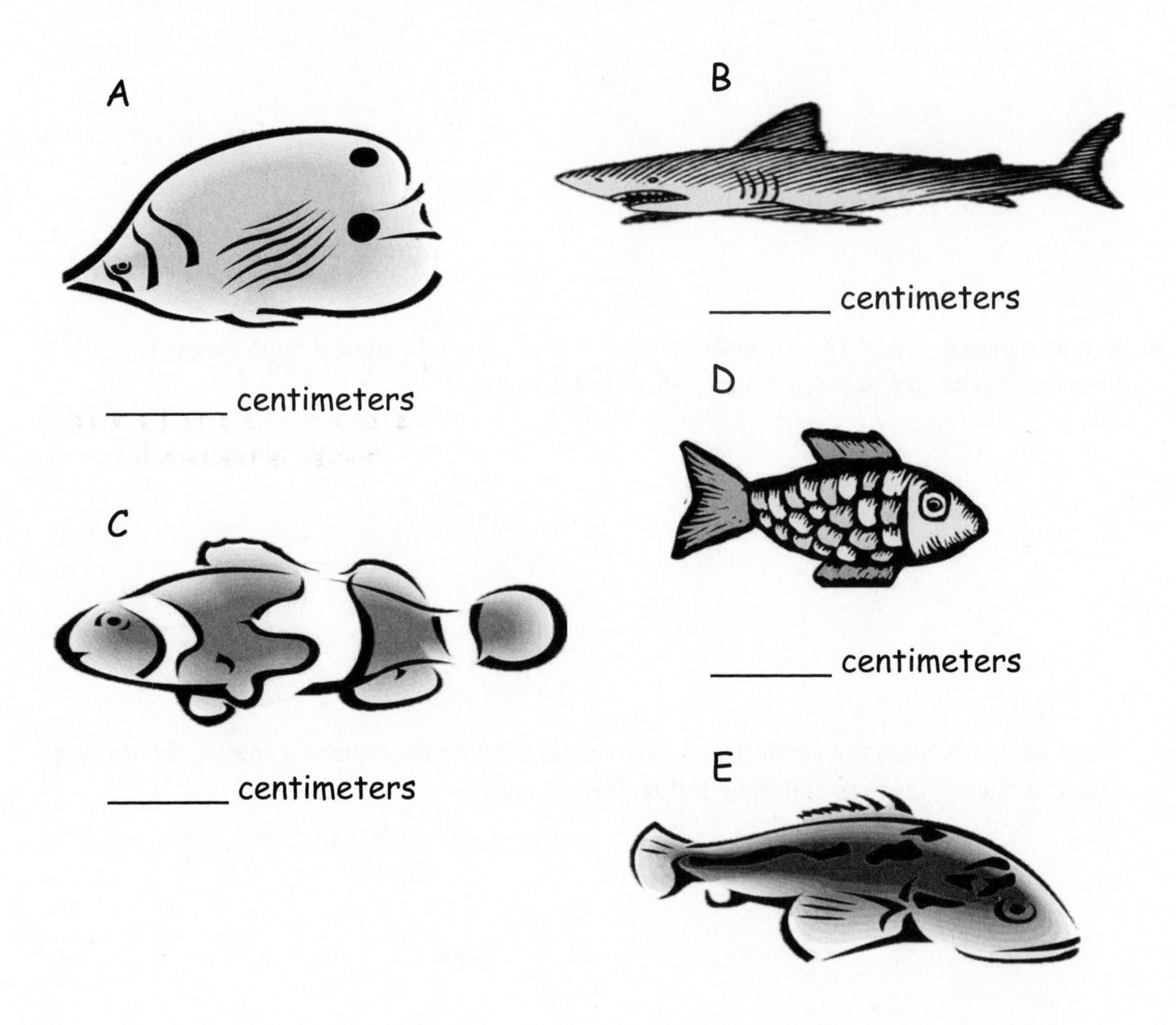

2. Order fish A, B, and C from longest to shortest.

__________ __________ __________

3. Use all of the fish measurements to complete the sentences.

   a. Fish A is longer than Fish _______ and shorter than Fish _______.

   b. Fish C is shorter than Fish _______ and longer than Fish _______.

   c. Fish _______ is the shortest fish.

   d. If Natasha gets a new fish that is shorter than Fish A, list the fish that the new fish is also shorter than.

Use your centimeter cubes to model each length, and answer the question.

4. Henry gets a new pencil that is 19 centimeters long. He sharpens the pencil several times. If the pencil is now 9 centimeters long, how much shorter is the pencil now than when it was new?

5. Malik and Jared each found a stick at the park. Malik found a stick that was 11 centimeters long. Jared found a stick that was 17 centimeters long. How much longer was Jared's stick?

This page intentionally left blank

Name ______________________________ Date ________________

1. Measure the length of each object with **LARGE** paper clips. Fill in the chart with your measurements.

Name of Object	Number of Large Paper Clips
a. bottle	
b. caterpillar	
c. key	
d. pen	
e. cow sticker	
f. Problem Set paper ↕	
g. reading book (from classroom)	

Cow

2. Measure the length of each object with **SMALL** paper clips. Fill in the chart with your measurements.

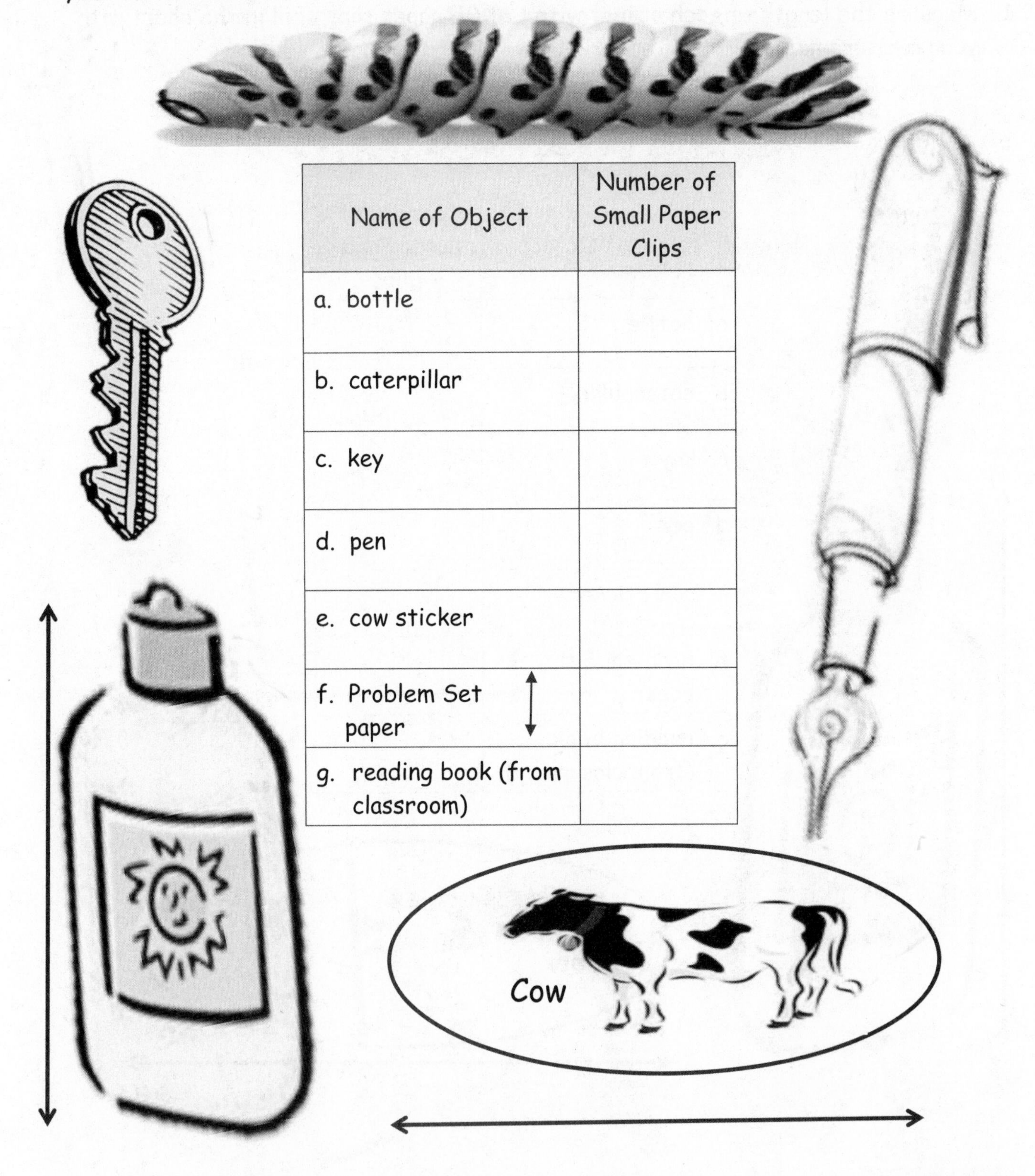

Name of Object	Number of Small Paper Clips
a. bottle	
b. caterpillar	
c. key	
d. pen	
e. cow sticker	
f. Problem Set paper ↕	
g. reading book (from classroom)	

Lesson 7: Measure the same objects from Topic B with different non-standard units simultaneously to see the need to measure with a consistent unit.

Name ____________________ Date __________

Cut the strip of paper clips. Measure the length of each object with the **large** paper clips to the right. Then, measure the length with the **small** paper clips on the back.

1. Fill in the chart on the back of the page with your measurements.

Paintbrush

Scissors

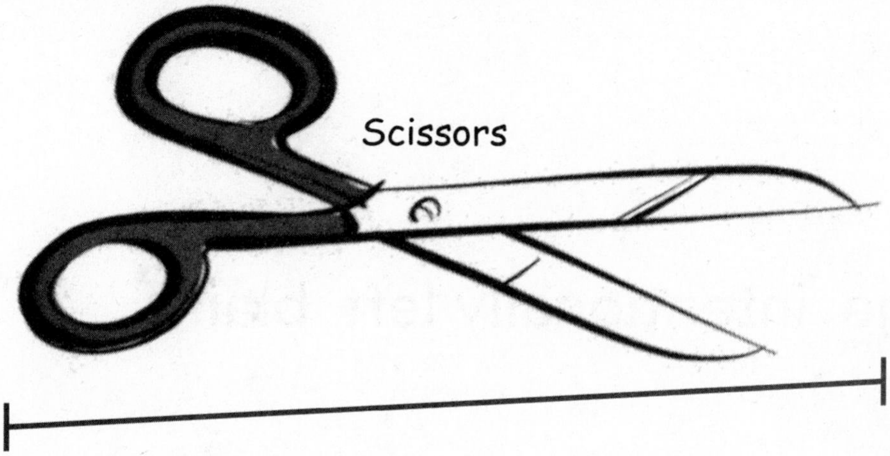

Glue

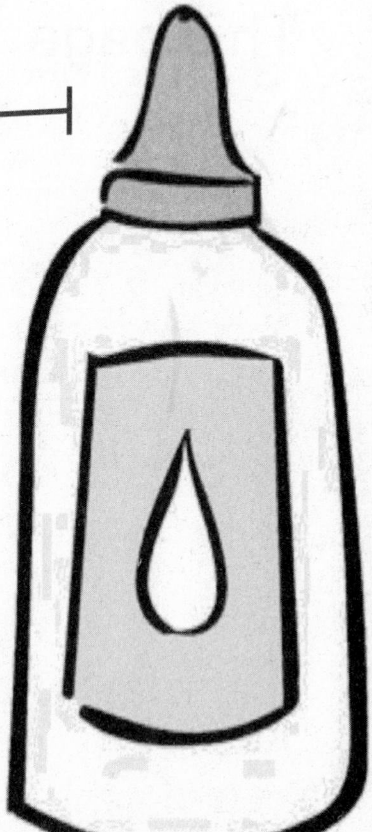

Crayon

Eraser

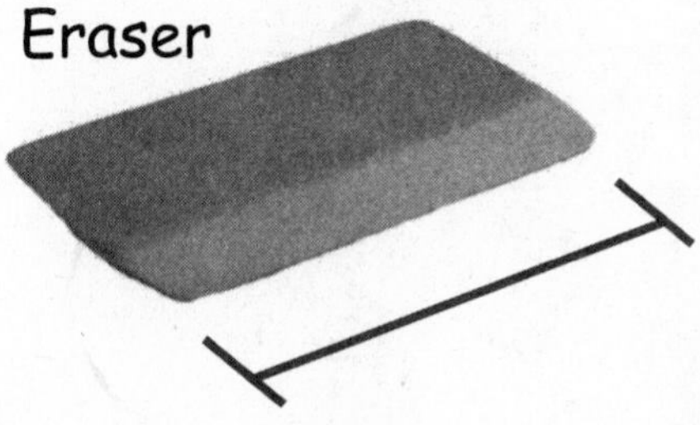

This page intentionally left blank

Name of Object	Length in Large Paper Clips	Length in Small Paper Clips
a. paintbrush		
b. scissors		
c. eraser		
d. crayon		
e. glue		

2. Find objects around your home to measure. Record the objects you find and their measurements on the chart.

Name of Object	Length in Large Paper Clips	Length in Small Paper Clips
a.		
b.		
c.		
d.		
e.		

This page intentionally left blank

Name ______________________________ Date ________________

Circle the length unit you will use to measure. Use the same length unit for all objects.

Small Paper Clips

Large Paper Clips

Toothpicks

Centimeter Cubes

Measure each object listed on the chart, and record the measurement. Add the names of other objects in the classroom, and record their measurements.

Classroom Object	Measurement
a. glue stick	
b. dry erase marker	
c. unsharpened pencil	
d. personal white board	
e.	
f.	
g.	

Name ______________________________ Date ________________

Circle the length unit you will use to measure. Use the same length unit for all objects.

Small Paper Clips

Large Paper Clips

Toothpicks

Centimeter Cubes

1. Measure each object listed on the chart, and record the measurement. Add the names of other objects in your house, and record their measurements.

Home Object	Measurement
a. fork	
b. picture frame	
c. pan	
d. shoe	

Home Object	Measurement
e. stuffed animal	
f.	
g.	

Did you remember to add the name of the length unit after the number? Yes No

2. Pick 3 items from the chart. List your items from longest to shortest:

a. ________________________

b. ________________________

c. ________________________

This page intentionally left blank

Name ______________________ Date ____________

1. Look at the picture below. How much **longer** is Guitar A than Guitar B?

GuitarA is ______ unit(s) **longer** than GuitarB.

2. Measure each object with centimeter cubes.

The blue pen is _____ ___________________.

The yellow pen is _____ _______________.

3. How much **longer** is the yellow pen than the blue pen?

   The yellow pen is _____ centimeters **longer** than the blue pen.

4. How much **shorter** is the blue pen than the yellow pen?

   The blue pen is _____ centimeters **shorter** than the yellow pen.

Use your centimeter cubes to model each problem. Then, solve by drawing a picture of your model and writing a number sentence and a statement.

5. Austin wants to make a train that is 13 centimeter cubes long. If his train is already 9 centimeter cubes long, how many **more** cubes does he need?

6. Kea's boat is 12 centimeters long, and Megan's boat is 8 centimeters long. How much **shorter** is Megan's boat than Kea's boat?

7. Kim cuts a piece of ribbon for her mom that is 14 centimeters long. Her mom says the ribbon is 8 centimeters too long. How **long** should the ribbon be?

8. The tail of Lee's dog is 15 centimeters long. If the tail of Kit's dog is 9 centimeters long, how much **longer** is the tail of Lee's dog than the tail of Kit's dog?

Name ____________________ Date ____________

1. Look at the picture below. How much **shorter** is Trophy A than Trophy B?

Trophy A is ______ units **shorter** than Trophy B.

2. Measure each object with centimeter cubes.

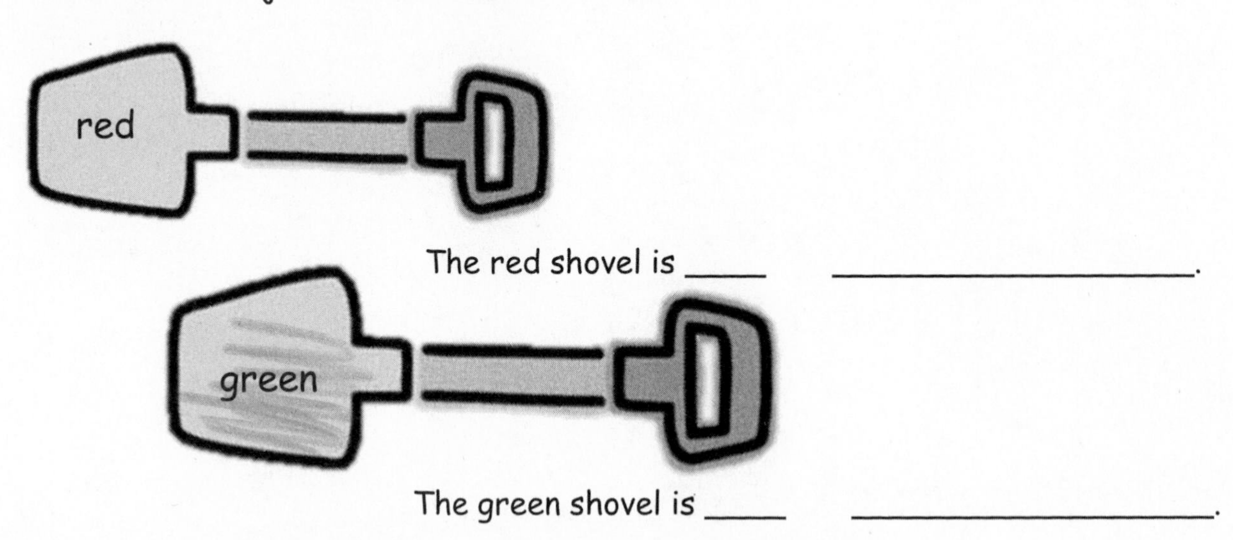

The red shovel is _____ ____________________.

The green shovel is _____ ____________________.

3. How much **longer** is the green shovel than the red shovel?
The green shovel is _____ centimeters **longer** than the red shovel.

Use your centimeter cubes to model each problem. Then, solve by drawing a picture of your model and writing a number sentence and a statement.

4. Susan grew 15 centimeters, and Tyler grew 11 centimeters. How much **more** did Susan grow than Tyler?

5. Bob's straw is 13 centimeters long. If Tom's straw is 6 centimeters long, how much **shorter** is Tom's straw than Bob's straw?

6. A purple card is 8 centimeters long. A red card is 12 centimeters long. How much **longer** is the red card than the purple card?

7. Carl's bean plant grew to be 9 centimeters tall. Dan's bean plant grew to be 14 centimeters tall. How much **taller** is Dan's plant than Carl's plant?

Name ______________________________ Date ________________

A group of people were asked to say their favorite color. Organize the data using tally marks, and answer the questions.

green red red blue red blue blue

red blue green blue red red

Red	
Green	
Blue	

1. How many people chose red as their favorite color? 6 people like red.

2. How many people chose blue as their favorite color? 5 people like blue.

3. How many people chose green as their favorite color? 2 people like green.

4. Which color received the least amount of votes? green

5. Write a number sentence that tells the total number of people who were asked their favorite color.

____

Name ______________________ Date ____________

Students were asked about their favorite ice cream flavor. Use the data below to answer the questions.

Ice Cream Flavor	Tally Marks	Votes
Chocolate	||||	
Strawberry	|||	
Cookie Dough	𝍸 𝍸	

1. Fill in the blanks in the table by writing the number of students who voted for each flavor.

2. How many students chose cookie dough as the flavor they like **best**?

   _____ students

3. What is the total number of students who like chocolate or strawberry the **best**?

   _____ students

4. Which flavor received the **least** amount of votes? ____________________

5. What is the total number of students who like cookie dough or chocolate the **best**?

   _____ students

6. Which two flavors were liked by a **total** of 7 students?

   ____________________ and ____________________

7. Write an addition sentence that shows how many students voted for their favorite ice cream flavor.

   ____

Students voted on what they like to read the most. Organize the data using tally marks, and then answer the questions.

comic book	magazine	chapter book	comic book	magazine
chapter book	comic book	comic book	chapter book	chapter book
chapter book	chapter book	magazine	magazine	magazine

What Students Like to Read the Most	Number of Students
Comic Book	
Magazine	
Chapter Book	

8. How many students like to read chapter books the most? _____ students

9. Which item received the **least** amount of votes? ________________

10. How many more students like to read chapter books than magazines?

    _____ students

11. What is the total number of students who like to read magazines or chapter books?

    _____ students

12. Which two items did a total of 9 students like to read?

    ______________________ and ______________________

13. Write an addition sentence that shows how many students voted.

    ____

This page intentionally left blank

Name ______________________________ Date ________________

Welcome to Data Day! Follow the directions to **collect** and **organize** data. Then, **ask** and **answer questions** about the data.

- Choose a question. Circle your choice.
- Pick 3 answer choices.
- Ask your classmates the question, and show them the 3 choices. Record the data on a class list.
- Organize the data in the chart below.

Which fruit do you like best?	Which snack do you like best?	What do you like to do on the playground the most?	Which school subject do you like the best?	Which animal would you most like to be?

Answer Choices	Number of Students
apple	

- Complete the question sentence frames to ask questions about your data.
- Trade papers with a partner, and have your partner answer your questions.

1. How many students liked ______________ the best?

2. Which category received the fewest votes? ______________

3. How many more students liked ______________ than ________________?

4. What is the total number of students who liked ______________ or ______________ the best?

5. How many students answered the question? How do you know?

Name ______________________________ Date ________________

Collect information about things you own. Use tally marks or numbers to organize the data in the chart below.

How many **pets** do you have?	How many **toothbrushes** are in your home?	How many **pillows** are in your home?	How many **jars of tomato sauce** are in your home?	How many **picture frames** are in your home?

- Complete the question sentence frames to ask questions about your data.
- Answer your own questions.

1. How many ____________ do you have? (Pick the item you have the **most** of.)

2. How many ____________ do you have? (Pick the item you have the **least** of.)

3. **Together**, how many picture frames and pillows do you have?

4. Write and answer two more questions using the data you collected.

   a. ____?

   b. ____?

Students voted on their favorite type of museum to visit. Each student could only vote once. Answer the questions based on the data in the table.

Science Museum	6 stick figures
Art Museum	8 stick figures
History Museum	6 stick figures

5. How many students chose art museums? ________ students

6. How many students chose the art museum or the science museum?

   ________ students

7. From this data, can you tell how many students are in this class? Explain your thinking.

Name _______________________ Date _____________

Use squares with no gaps or overlaps to organize the data from the picture. Line up your **squares** carefully.

Favorite Ice Cream Flavor □ = 1 student

Number of Students

Flavors	
□ vanilla	
■ chocolate	

1. How many **more** students liked chocolate than liked vanilla? __________ students

2. How many **total** students were asked about their favorite ice cream flavor?

__________ students

Ties on Shoes Number of Students □ = 1 student

Types of Shoe Ties	
Velcro	□□□□□
laces	□□□□□□□□
no ties	□□□□□□□

3. Write a number sentence to show how many **total** students were asked about their shoes.

____

4. Write a number sentence to show how many **fewer** students have Velcro on their shoes than laces.

____

Each student in the class added a sticky note to show his or her favorite kind of pet. Use the graph to answer the questions.

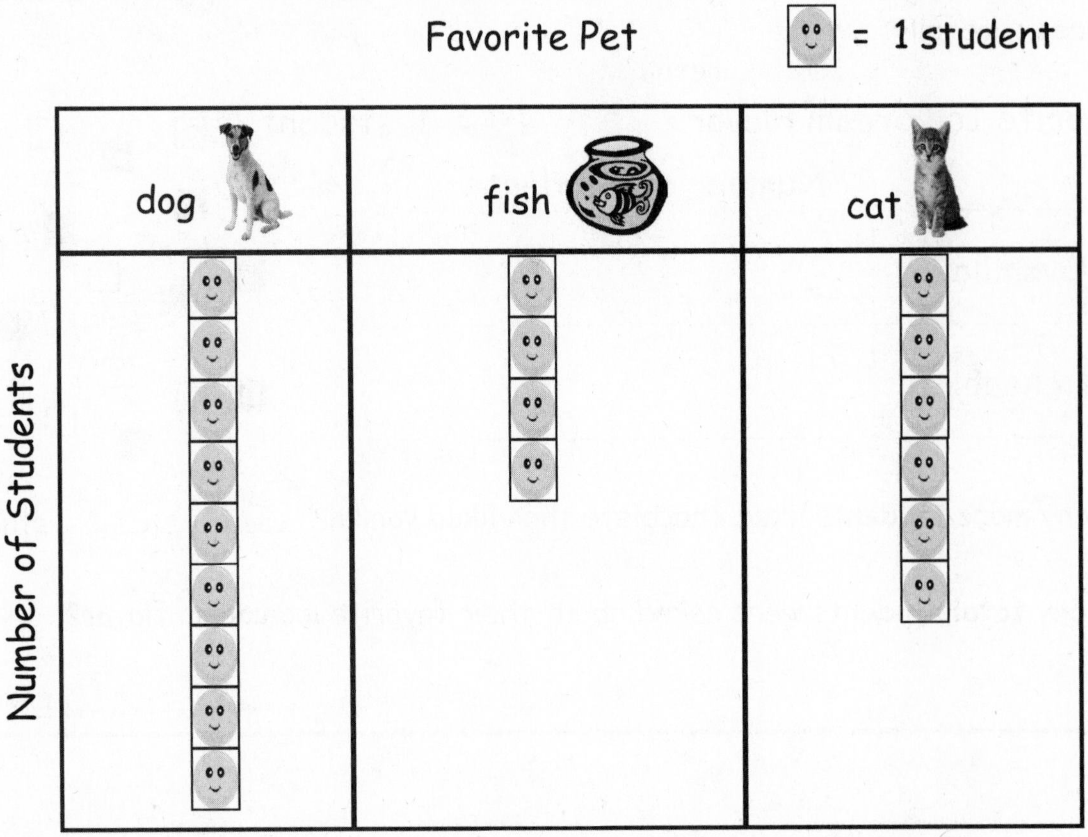

Favorite Pet = 1 student

dog	fish	cat

Number of Students

5. How many students chose dogs or cats as their favorite pet?

   ________ students

6. How many more students chose dogs as their favorite pet than cats?

   ________ students

7. How many more students chose cats than fish?

   ________ students

Name ______________________________ Date ________________

The class has 18 students. On Friday, 9 students wore sneakers, 6 students wore sandals, and 3 students wore boots. Use squares with no gaps or overlaps to organize the data. Line up your **squares** carefully.

Shoes Worn on Friday   Number of Students   □ = 1 student

Shoes	

1. How many more students wore sneakers than sandals? ________ students

2. Write a number sentence to tell how many students were asked about their shoes on Friday.

   ____

3. Write a number sentence to show how many fewer students wore boots than sneakers.

   ____

Our school garden has been growing for two months. The graph below shows the numbers of each vegetable that have been harvested so far.

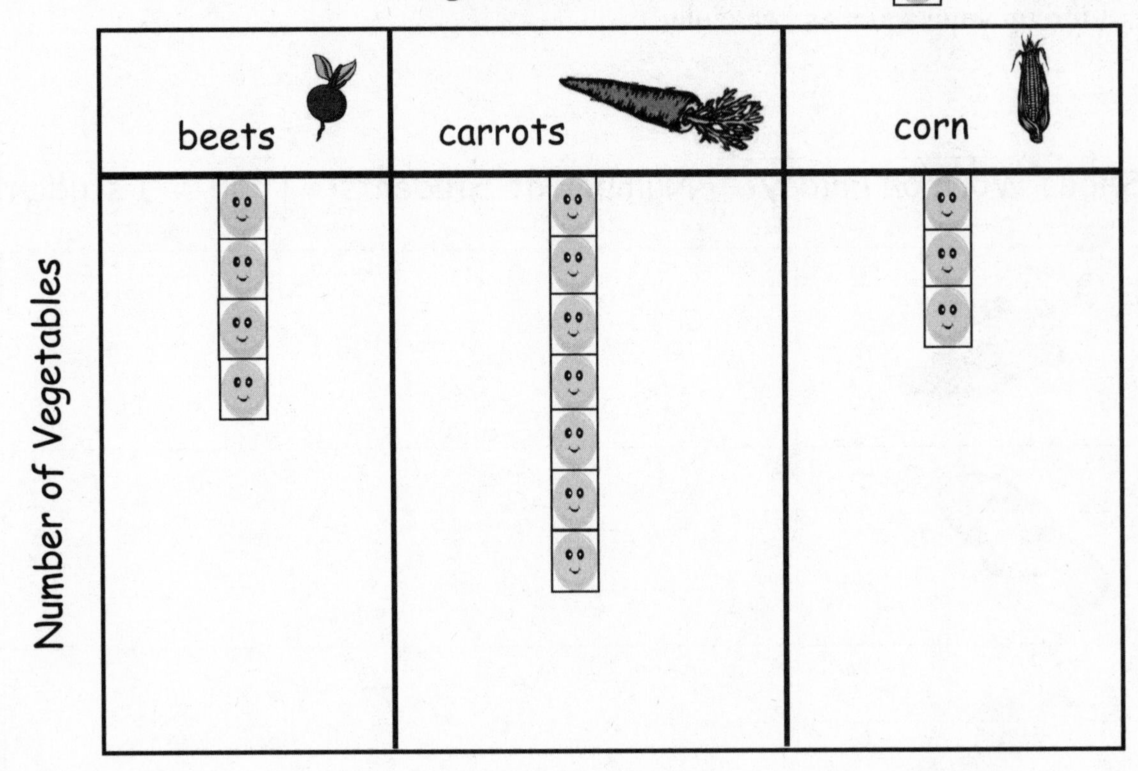

4. How many total vegetables were harvested?

_______ vegetables

5. Which vegetable has been harvested the most?

____

6. How many more beets were harvested than corn?

_______ more beets than corn

7. How many more beets would need to be harvested to have the same amount as the number of carrots harvested?

____

Name ____________________ Date __________

Use the graph to answer the questions. Fill in the blank, and write a number sentence to the right to solve the problem.

School Day Weather   ☐ = 1 day

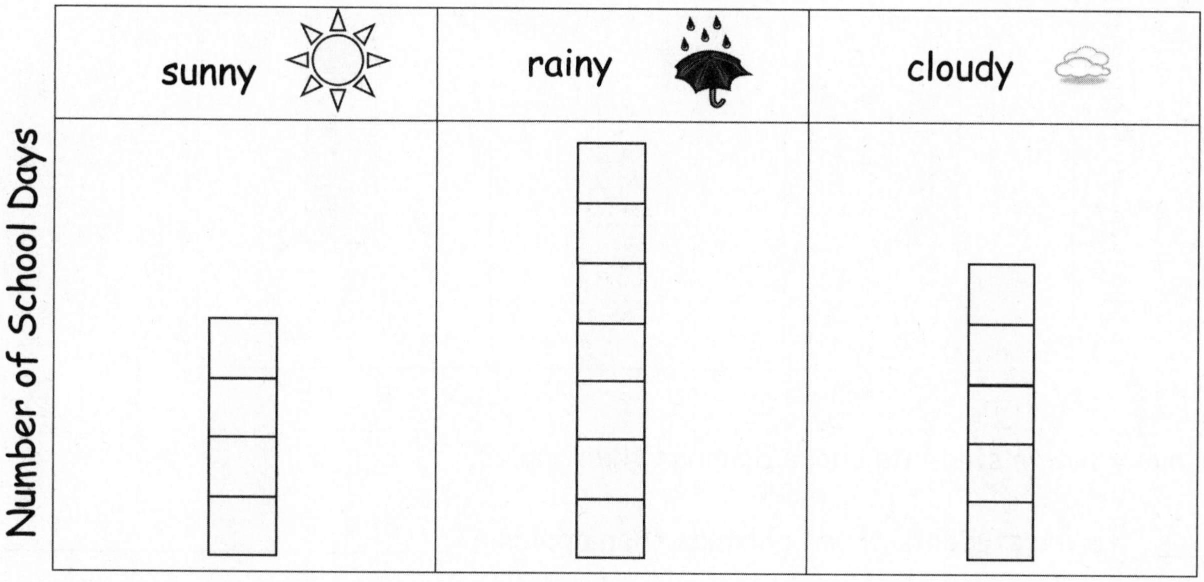

	sunny	rainy	cloudy
Number of School Days	☐☐☐☐	☐☐☐☐☐☐☐	☐☐☐☐☐

1. How many more days were cloudy than sunny?

   _______ more day(s) were cloudy than sunny. ____________________

2. How many fewer days were cloudy than rainy?

   _______ more day(s) were cloudy than rainy. ____________________

3. How many more days were rainy than sunny?

   _______ more day(s) were rainy than sunny. ____________________

4. How many total days did the class keep track of the weather?

   The class kept track of a total of _______ days. ____________________

5. If the next 3 school days are sunny, how many of the school days will be sunny in all?

   _______ days will be sunny. ____________________

Use the graph to answer the questions. Fill in the blank, and write a number sentence that helps you solve the problem.

Favorite Fruit (☺ = 1 student)

Number of Students

Apple	Banana	Grapes
☺☺☺☺☺☺	☺☺☺☺☺	☺☺☺☺

6. How many fewer students chose bananas than apples?

   _______ fewer students chose bananas than apples. ____________________

7. How many more students chose bananas than grapes?

   _______ more students chose bananas than grapes. ____________________

8. How many fewer students chose grapes than apples?

   _______ fewer students chose grapes than apples. ____________________

9. Some more students answered about their favorite fruits. If the new total number of students who answered is 20, how many more students answered?

   _______ more students answered the question. ____________________

EUREKA MATH™

Name ______________________________ Date ________________

Use the graph to answer the questions. Fill in the blank, and write a number sentence.

School Lunch Order = 1 student

hot lunch	sandwich	salad

1. How many more hot lunch orders were there than sandwich orders?

   There were _____ more hot lunch orders.

   ____

2. How many fewer salad orders were there than hot lunch orders?

   There were _____ fewer salad orders.

   ____

3. If 5 more students order hot lunch, how many hot lunch orders will there be?

   There will be _____ hot lunch orders.

   ____

Use the table to answer the questions. Fill in the blanks, and write a number sentence.

Favorite Type of Book

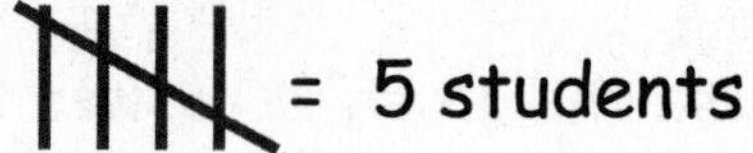
= 5 students

fairy tales	𝍸 𝍸 |
science books	𝍸 |||
poetry books	𝍸 𝍸 𝍸

4. How many more students like fairy tales than science books?

   ______ more students like fairy tales. ____________________

5. How many fewer students like science books than poetry books?

   ______ fewer students like science books. ____________________

6. How many students picked fairy tales or science books in all?

   ______ students picked fairy tales or science books. ____________________

7. How many more students would need to pick science books to have the same number of books as fairy tales?

   ______ more students would need to pick science books. ____________________

8. If 5 more students show up late and all pick fairy tales, will this be the most popular book? Use a number sentence to show your answer.

   ____

Student Edition

Eureka Math
Grade 1
Module 4

Special thanks go to the Gordon A. Cain Center and to the Department of Mathematics at Louisiana State University for their support in the development of *Eureka Math*.

For a free *Eureka Math* Teacher Resource Pack, Parent Tip Sheets, and more please visit www.Eureka.tools

Published by the non-profit Great Minds

Printed in the U.S.A.
This book may be purchased from the publisher at eureka-math.org
10 9

Name ______________________ Date ____________

Write the tens and ones and complete the statement.

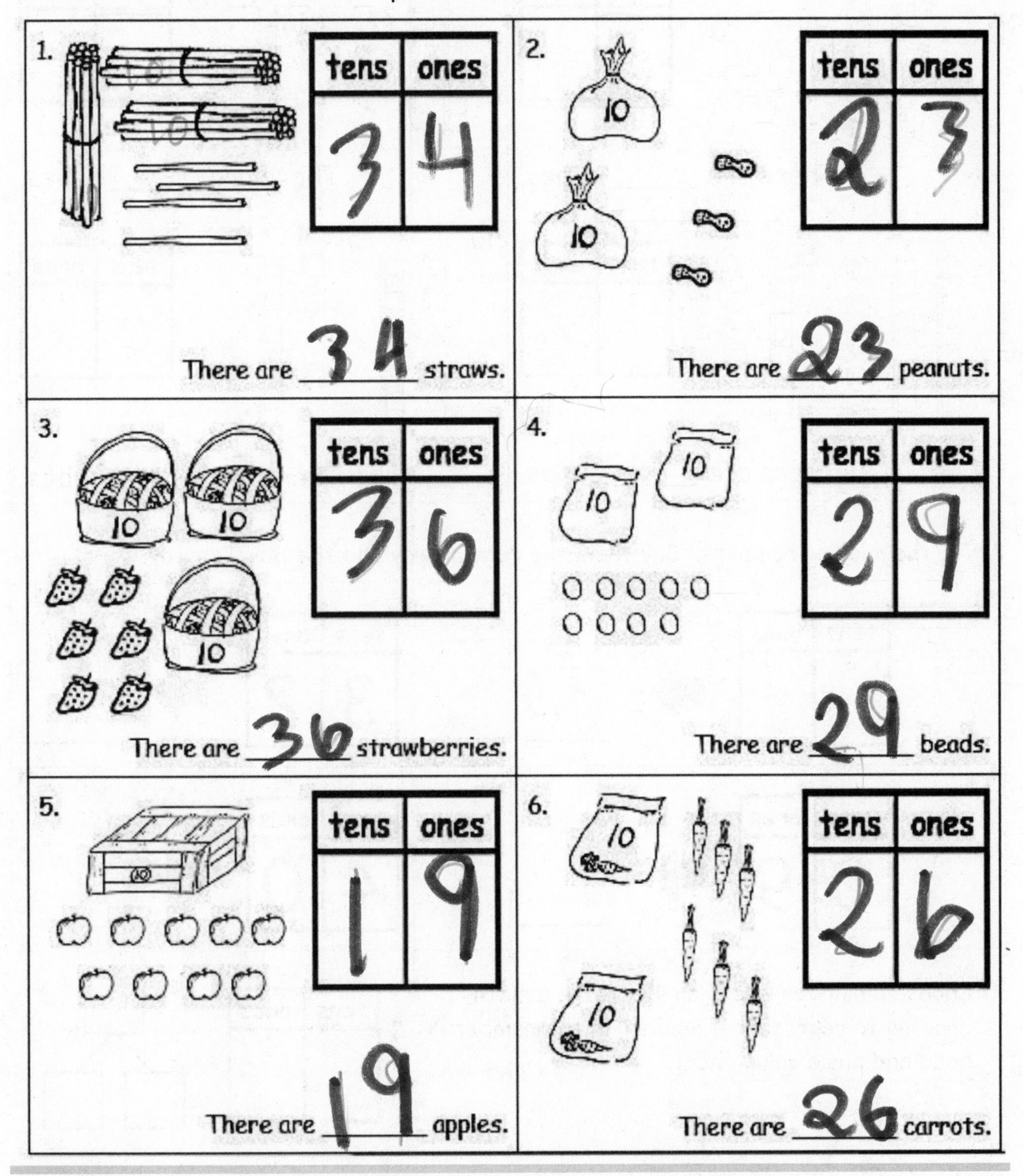

Write the tens and ones. Complete the statement.

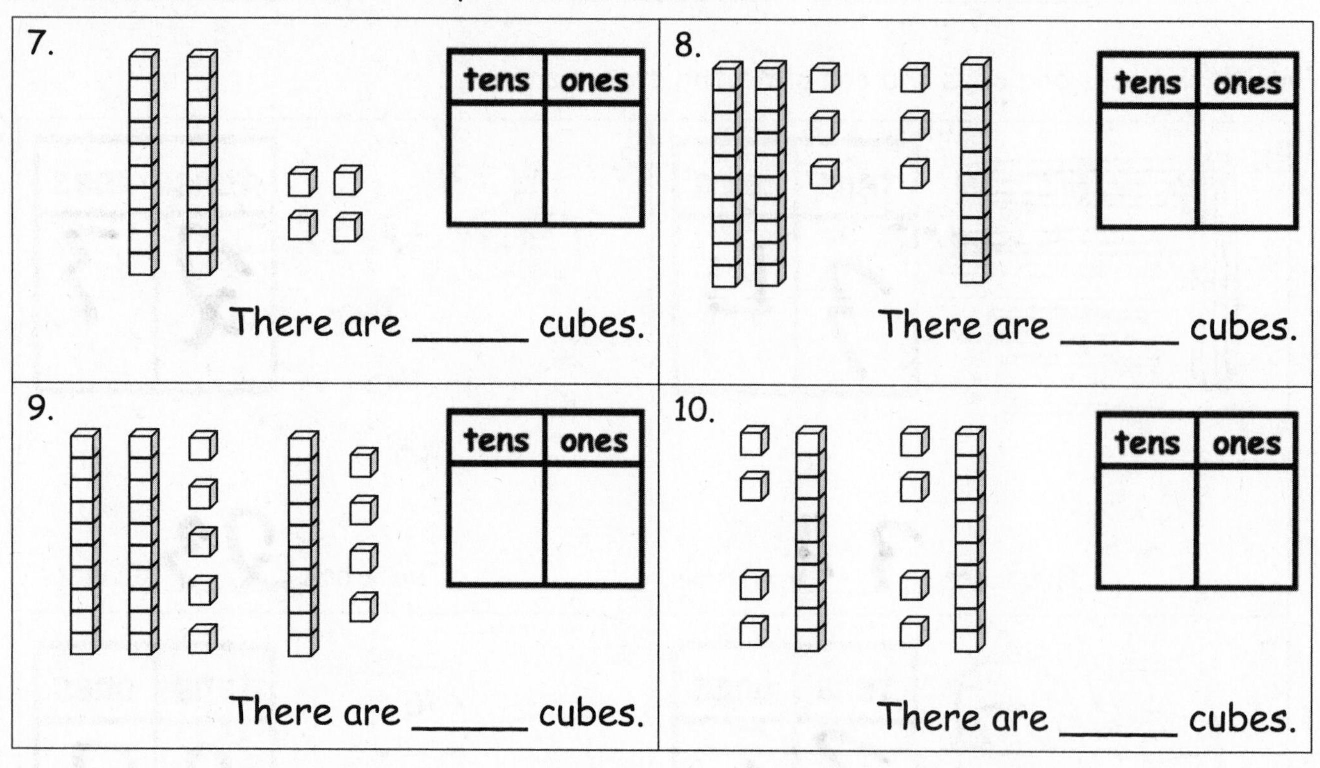

Write the missing numbers. Say them the regular way and the Say Ten way.

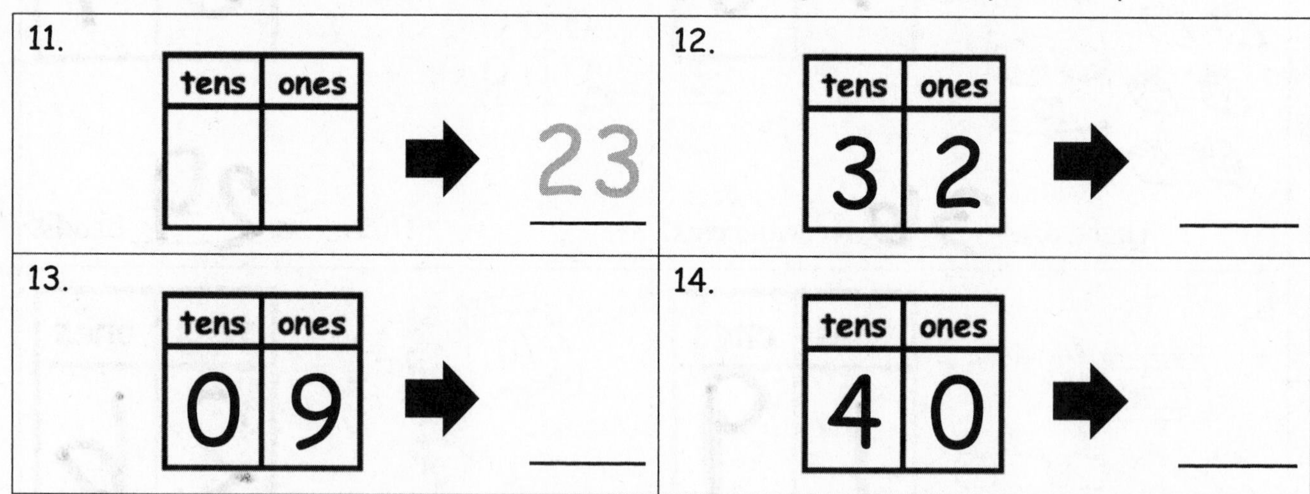

15. Choose a number less than 40. Make a math drawing to represent it, and fill in the number bond and place value chart.

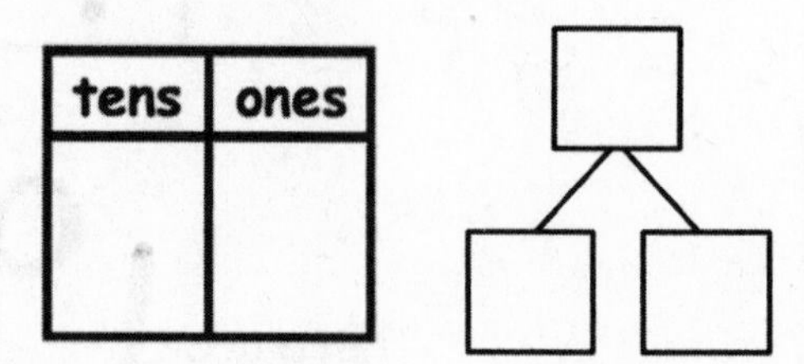

EUREKA MATH

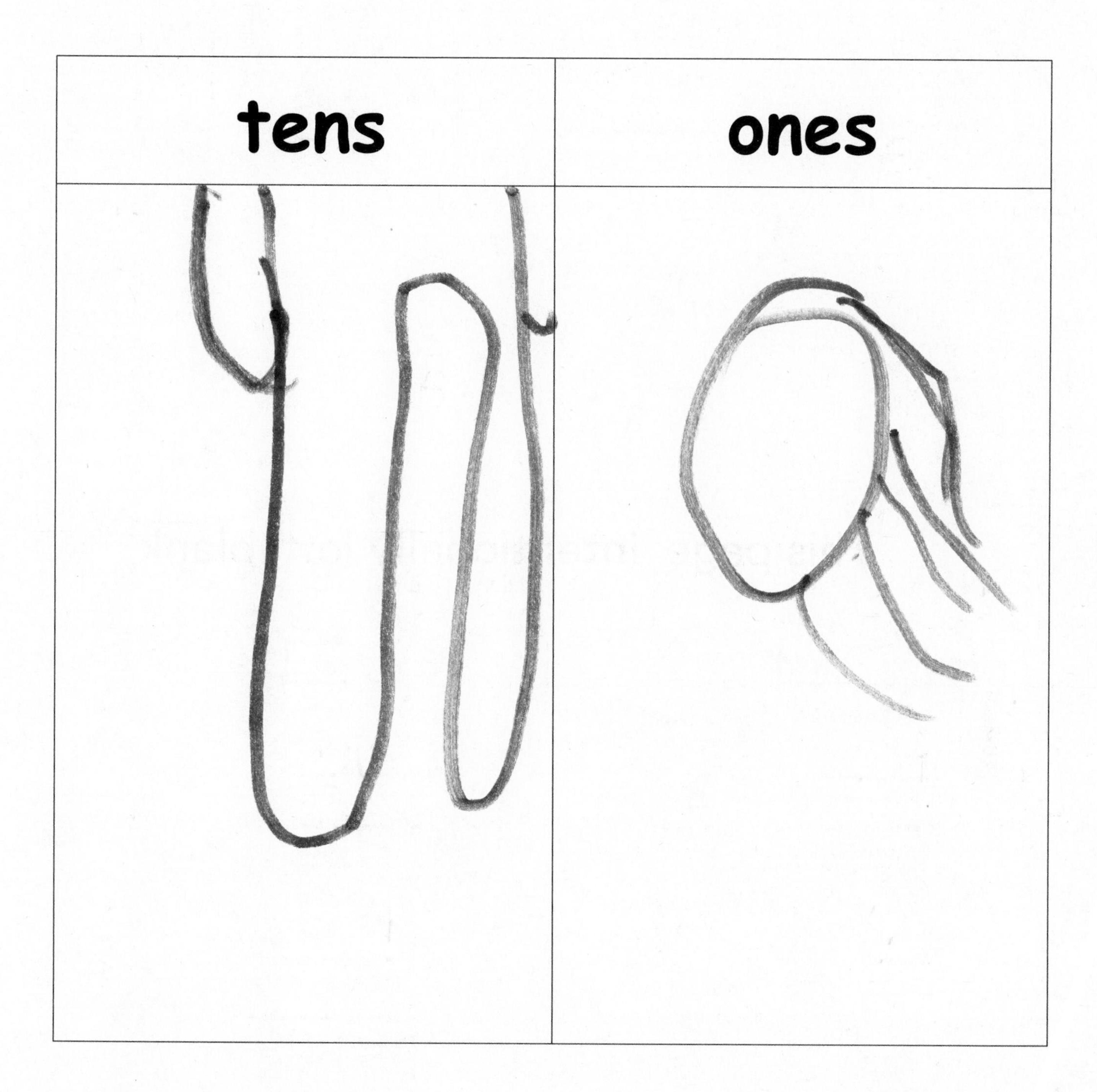

place value chart

This page intentionally left blank

Name ________________________ Date ______________

Count as many tens as you can. Complete each statement. Say the numbers and the sentences.

1.

_____ ten _____ ones is the

same as ______ ones.

2.

_____ tens _____ ones is the

same as ______ ones.

3.

_____ tens _____ ones is the

same as ______ ones.

4.

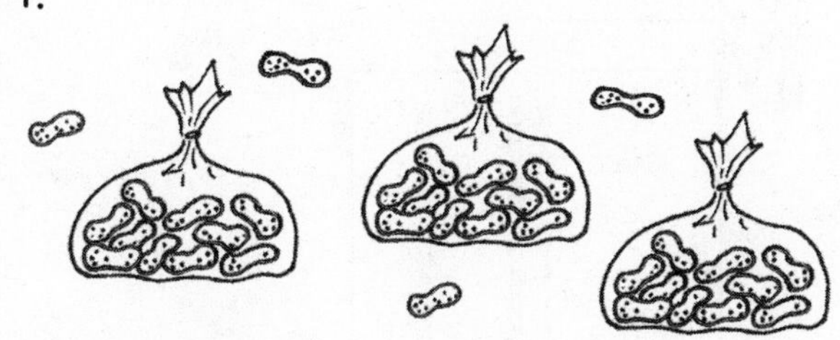

_____ tens _____ ones is the

same as ______ ones.

5.

_____ tens _____ ones is the

same as ______ ones.

6.

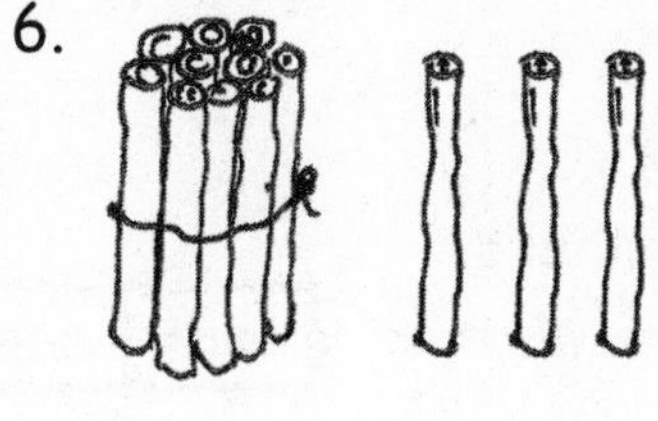

_____ ten _____ ones is the

same as ______ ones.

Match.

7. 3 tens 2 ones

8.

tens	ones
1	7

9. 37 ones

10. 4 tens

11.

12. 9 ones 2 tens

29 ones

40 ones

23 ones

32 ones

17 ones

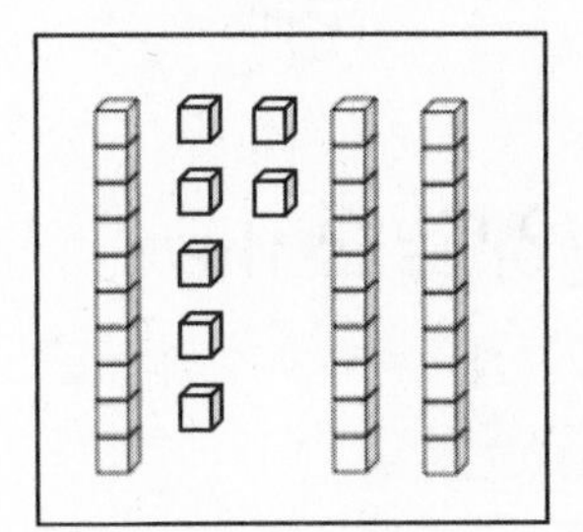

Fill in the missing numbers.

13. 15 →

tens	ones

→ ______ ones

14. ______  ______ tens ______ ones  39 ones

Name ______________________ Date ____________

Count as many tens as you can. Complete each statement. Say the numbers and the sentences.

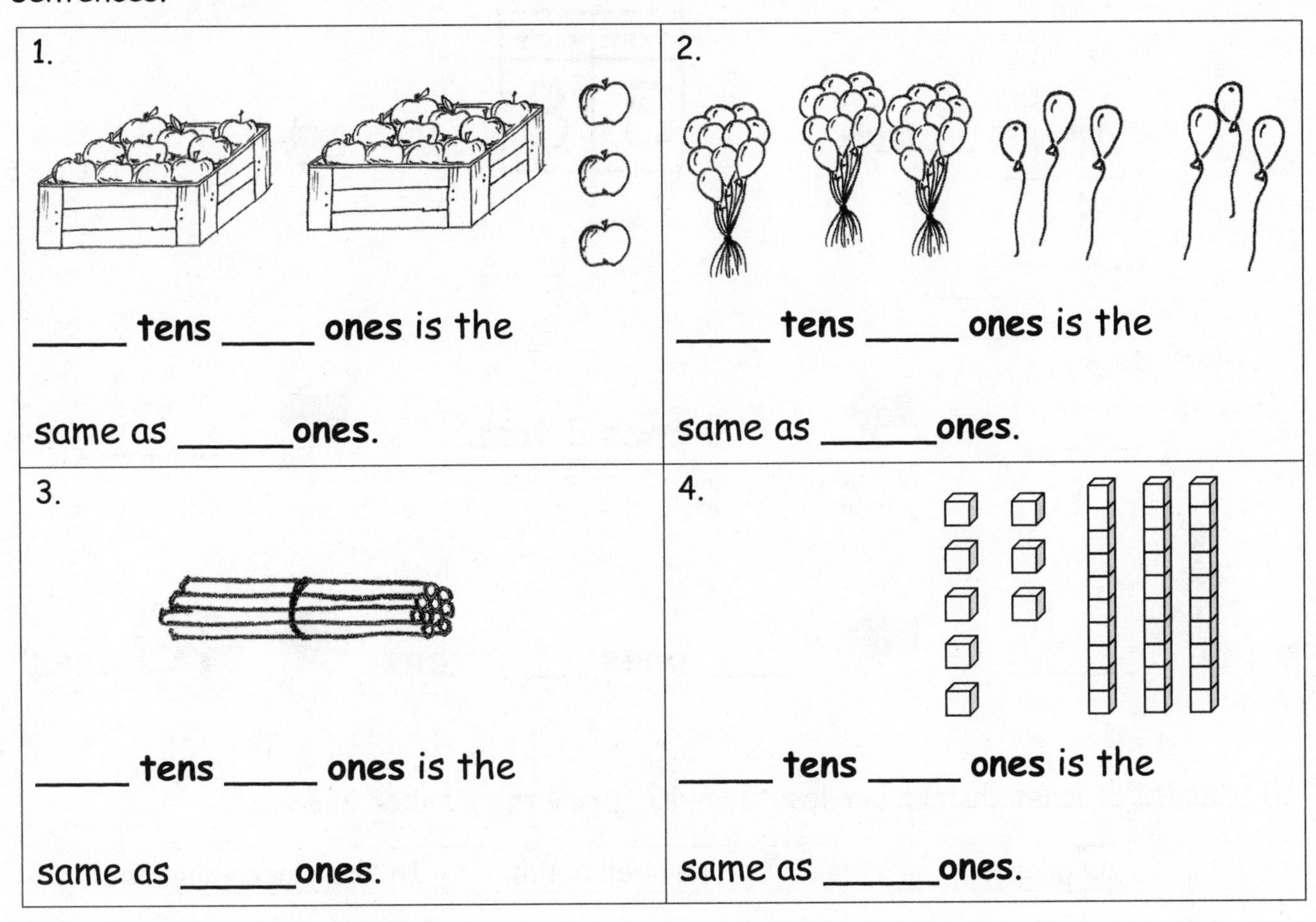

1. ____ tens ____ ones is the same as _____ones.

2. ____ tens ____ ones is the same as _____ones.

3. ____ tens ____ ones is the same as _____ones.

4. ____ tens ____ ones is the same as _____ones.

Fill in the missing numbers.

5. ______ ➡

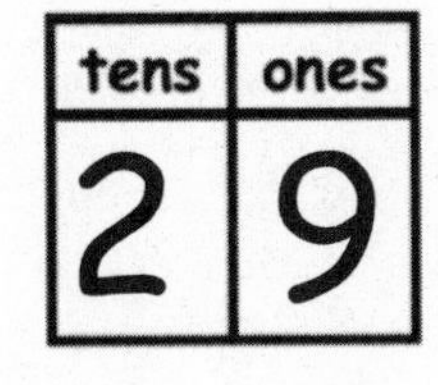

➡

6. 34 ➡ ____ tens ____ ones ➡ ______ ones

7. ______ ➡ [tens 3 | ones 8] ➡ ______ ones

8. ______ ➡ 9 ones 3 tens ➡ ______ ones

9. ______ ➡ ____ ones ____ tens ➡ 40 ones

10. Choose at least one number less than 40. Draw the number in 3 ways:

As grapes:	In a number bond:	In the place value chart:
		tens / ones

Name ____________________ Date __________

Write the number.

1.

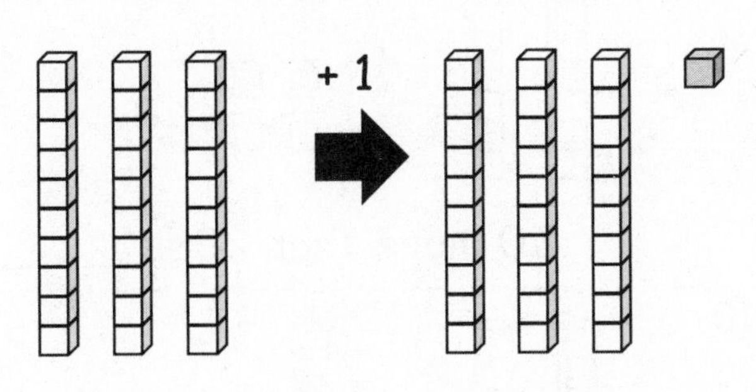

1 more than 30 is ______.

2.

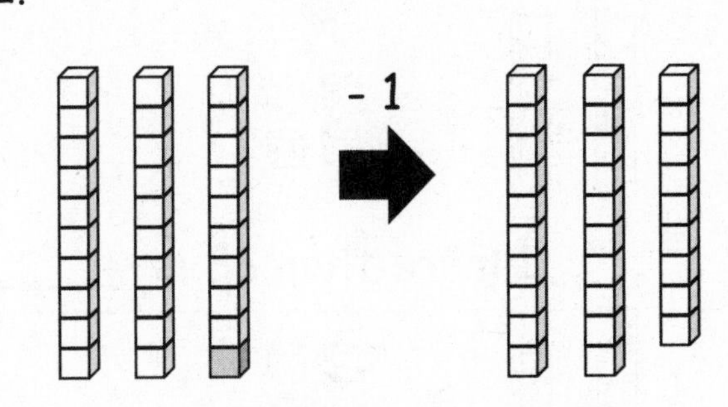

1 less than 30 is ______.

3.

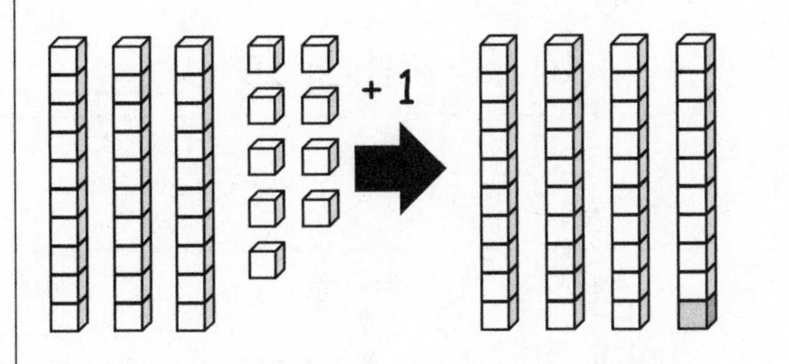

1 more than 39 is ______.

4.

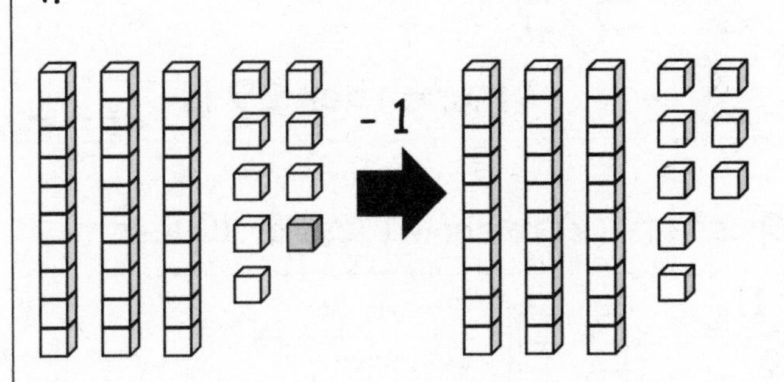

1 less than 39 is ______.

5.

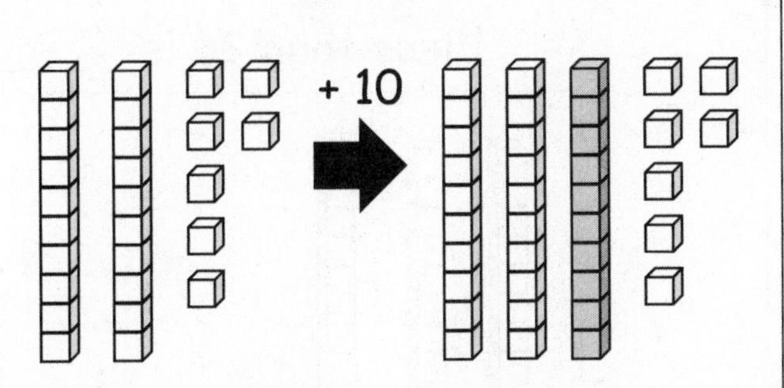

10 more than 27 is ______.

6.

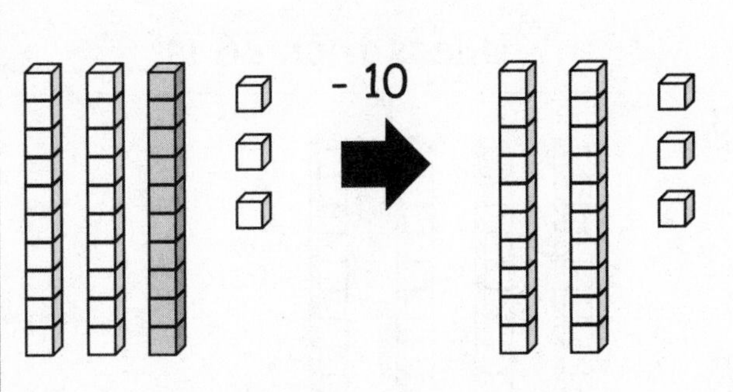

10 less than 33 is ______.

Draw 1 more or 10 more. You may use a quick ten to show 10 more.

7. 1 more than 28 is ______.	8. 10 more than 28 is ______.
9. 1 more than 29 is ______.	10. 10 more than 29 is ______.

Cross off (x) to show 1 less or 10 less.

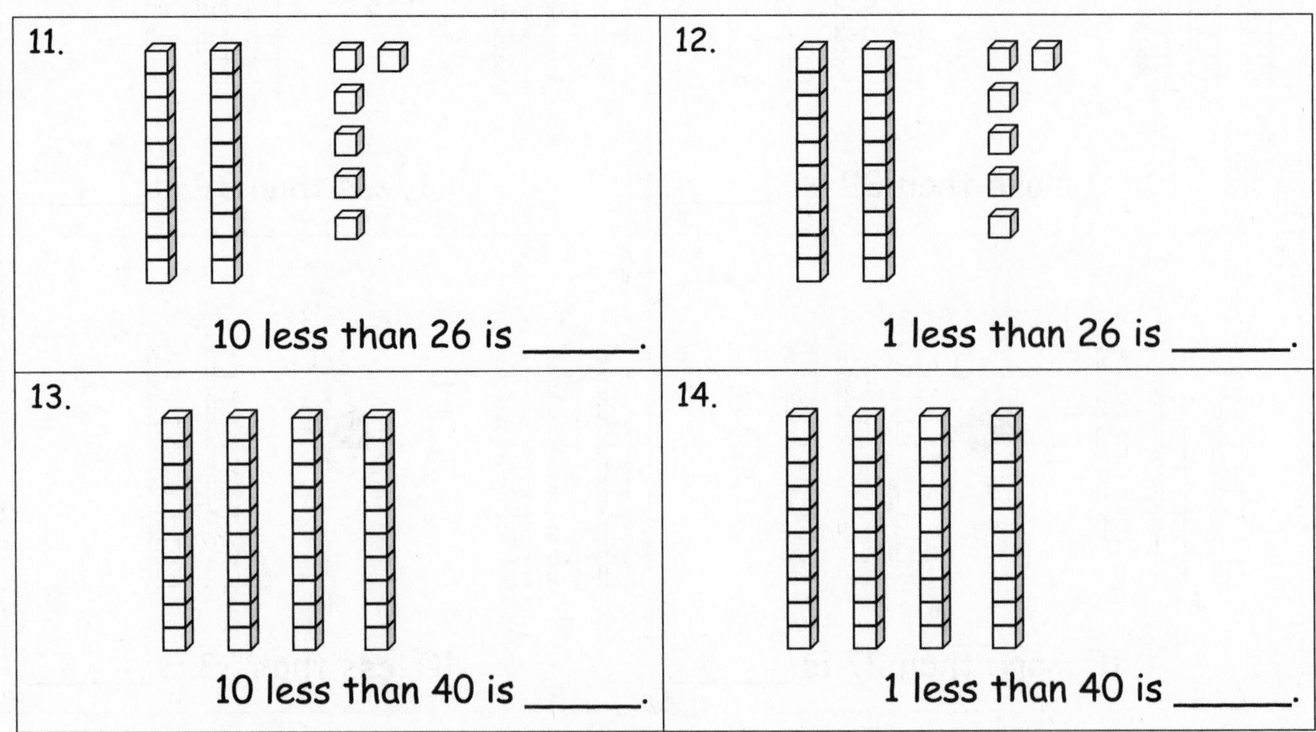

Name ______________________________ Date ______________

Draw quick tens and ones to show the number. Then, draw 1 more or 10 more.

1. 1 more than 38 is _____.	2. 10 more than 38 is _____.
3. 1 more than 35 is _____.	4. 10 more than 35 is _____.

Draw quick tens and ones to show the number. Cross off (x) to show 1 less or 10 less.

5. 10 less than 23 is _____.	6. 1 less than 23 is _____.
7. 10 less than 31 is _____.	8. 1 less than 31 is _____.

Match the words to the picture that shows the right amount.

9.

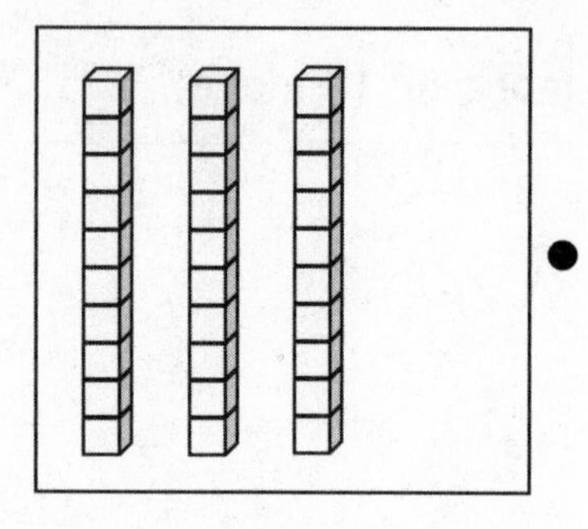

• • 1 less than 30.

10.

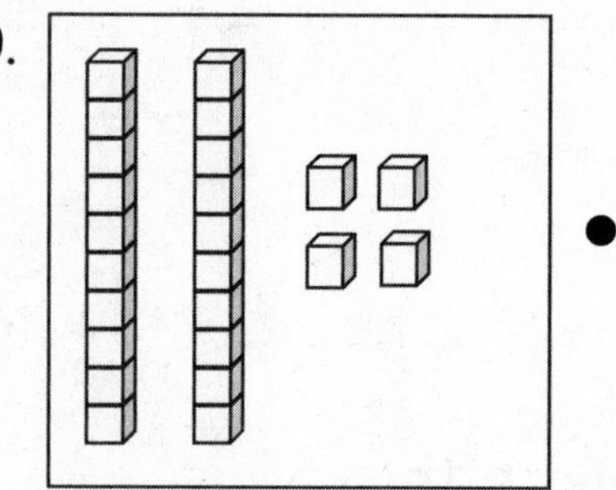

• • 1 more than 23.

11.

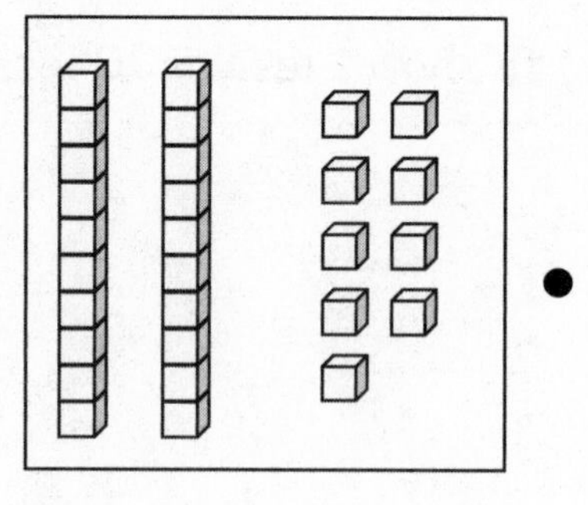

• • 10 less than 36.

12.

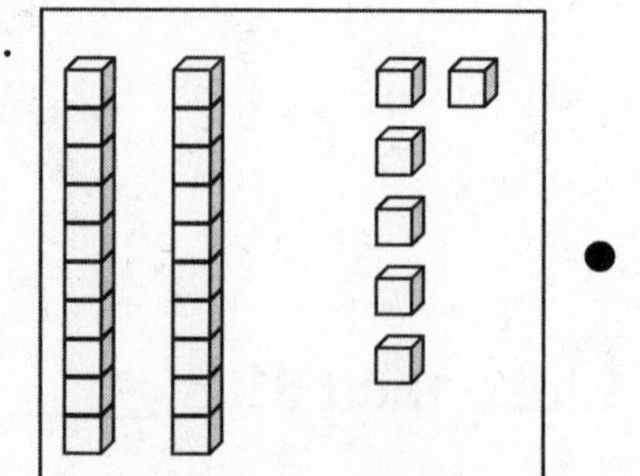

• • 10 more than 20.

tens	ones

tens	ones

double place value charts

This page intentionally left blank

Name ______________________________ Date ______________

Fill in the place value chart and the blanks.

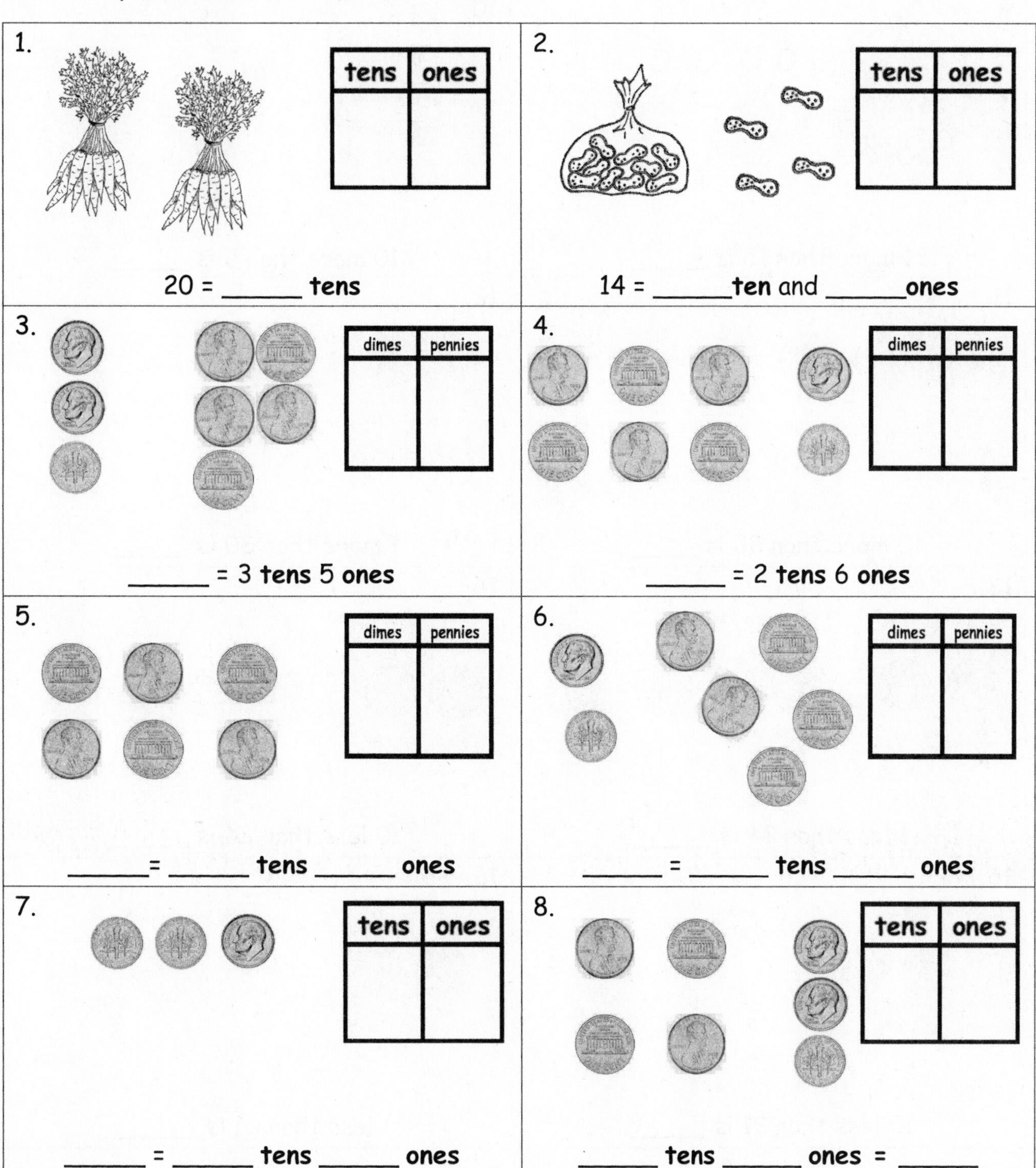

Fill in the blank. Draw or cross off tens or ones as needed.

10 more than 25 is 35

9.

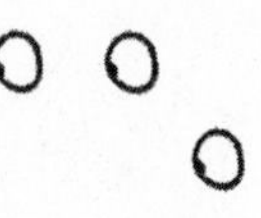

1 more than 15 is ______.

10.

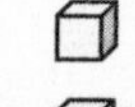

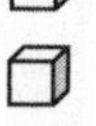

10 more than 5 is ______.

11.

10 more than 30 is ______.

12.

1 more than 30 is ______.

13.

1 less than 24 is ______.

14.

10 less than 24 is ______.

15.

10 less than 21 is ______.

16.

1 less than 21 is ______.

Name _______________________ Date ____________

Fill in the place value chart and the blanks.

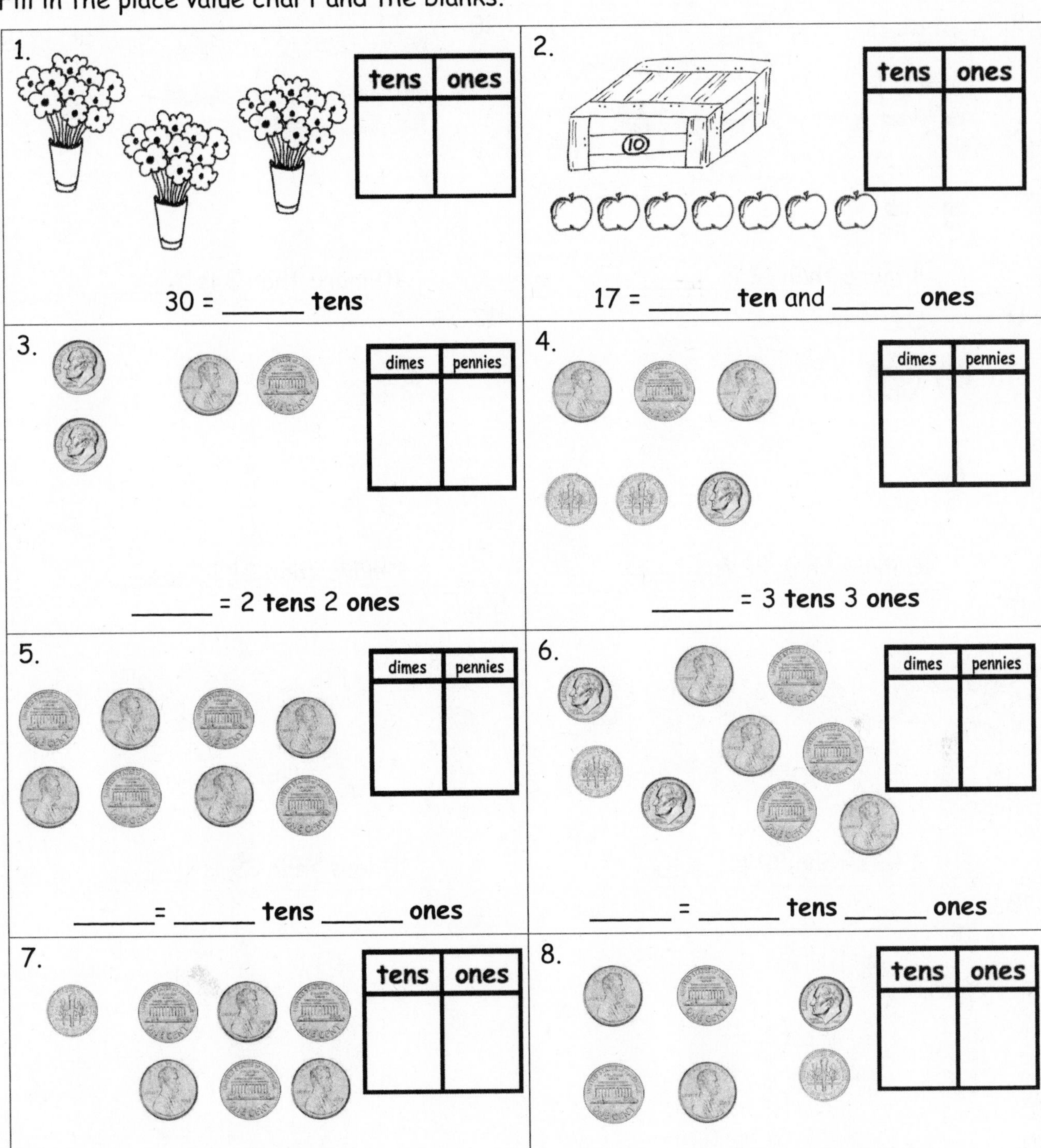

Fill in the blank. Draw or cross off tens or ones as needed.

9. 1 more than 12 is 13.

10. 10 more than 3 is 13.

11. 10 more than 22 is ______.

12. 1 more than 22 is ______.

13. 1 less than 39 is ______.

14. 10 less than 39 is ______.

15. 10 less than 33 is ______.

16. 1 less than 33 is ______.

dimes	pennies

tens	ones

coin and place value charts

This page intentionally left blank

Name Liz Chatter Date ________________

For each pair, write the number of items in each set. Then, circle the set with the *greater* number of items.

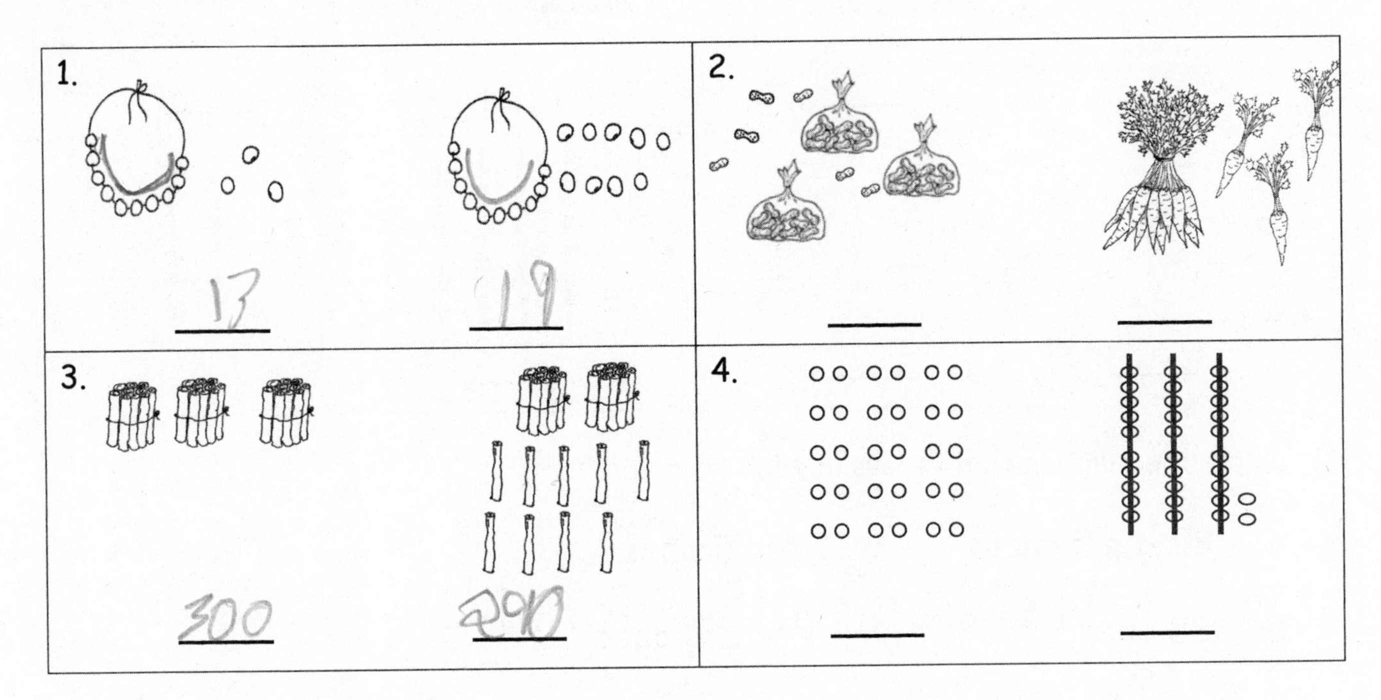

5. Circle the number that is *greater* in each pair.

a. 1 ten 2 ones 3 tens 2 ones

b. 2 tens 8 ones 3 tens 2 ones

c. 19 15

d. 31 26

6. Circle the set of coins that has a *greater* value.

3 dimes

3 pennies

For each pair, write the number of items in each set. Circle the set with *fewer* items.

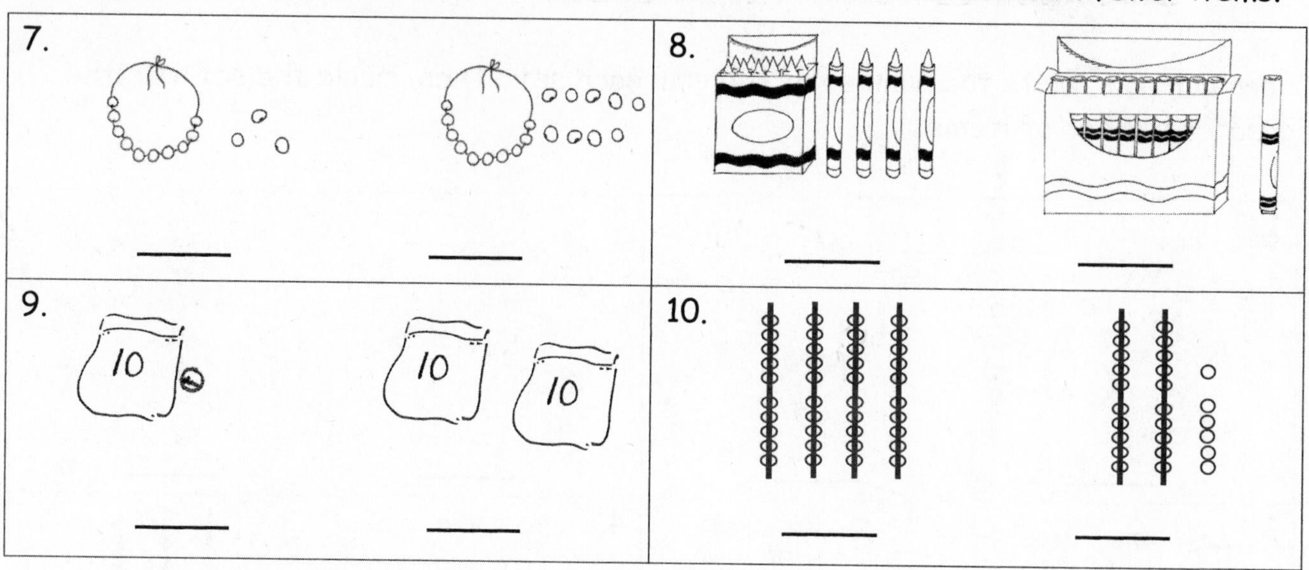

11. Circle the number that is *less* in each pair.

a. 2 tens 5 ones 1 ten 5 ones

b. 28 ones 3 tens 2 ones

c. 18 13

d. 31 26

12. Circle the set of coins that has *less* value.

1 dime 2 pennies

1 penny 2 dimes

13. Circle the amount that is *less*. Draw or write to show how you know.

32 17

Name ____________________ Date ____________

Write the number, and circle the set that is *greater* in each pair. Say a statement to compare the two sets.

1.

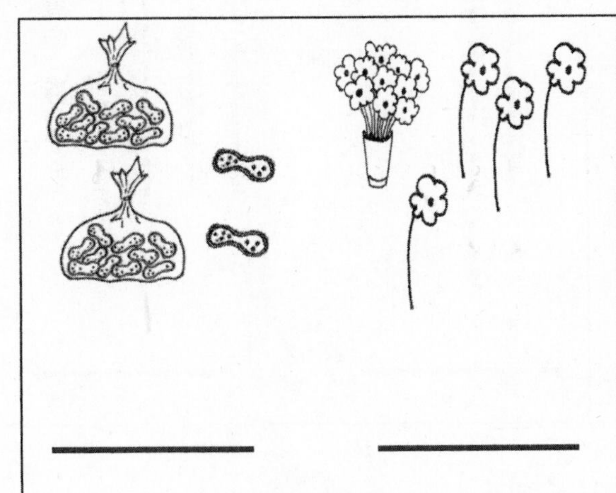

2.

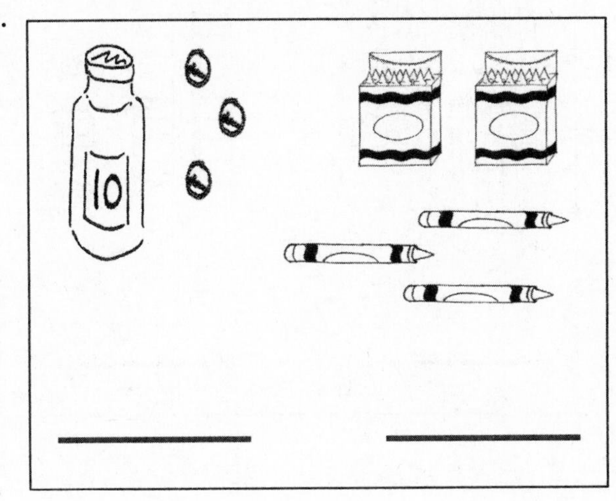

Circle the number that is *greater* for each pair.

3.

3 tens 8 ones	3 tens 9 ones

4.

25	35

5. Write the value and circle the set of coins that has *greater* value.

____

____

Write the number, and circle the set that is *less* in each pair. Say a statement to compare the two sets.

6.

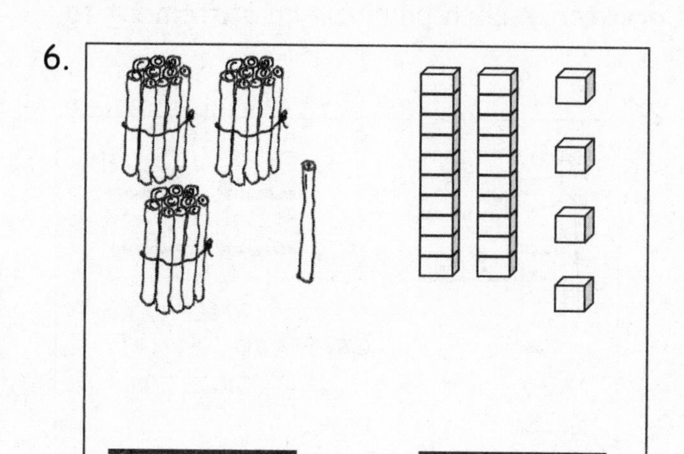

7.

Circle the number that is *less* for each pair.

8.

2 tens 7 ones	3 tens 7 ones

9.

22	29

10. Write the value and circle the set of coins that has *less* value.

11. Katelyn and Johnny are playing comparison with cards. They have recorded the totals for each round. For each round, circle the total that won the cards, and write the statement. The first one is done for you.

ROUND 1: The total that is **greater** wins.

Katelyn's Total
16

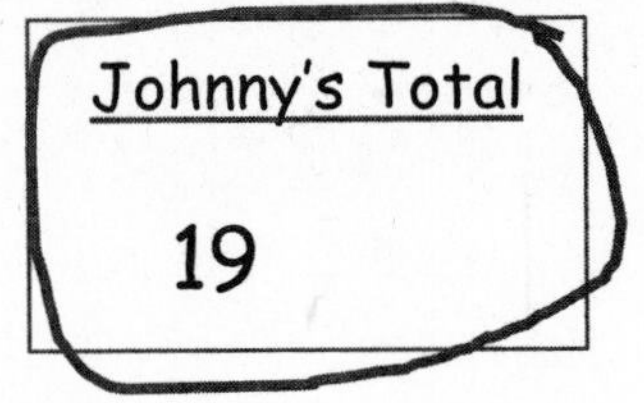

19 is greater than 16.

a. ROUND 2: The total that is **less** wins.

Katelyn's Total	Johnny's Total
27	24

____

b. ROUND 3: The total that is **greater** wins.

Katelyn's Total	Johnny's Total
32	22

____

c. ROUND 4: The total that is **less** wins.

Katelyn's Total	Johnny's Total
29	26

____

d. If Katelyn's total is 39, and Johnny's total has 3 tens 9 ones, who would have a greater total? Draw a math drawing to explain how you know.

This page intentionally left blank

Name ______________________ Date ____________

Word Bank

is greater than
is less than
is equal to

1. Draw quick tens and ones to show each number. Label the first drawing as *less than (L)*, *greater than (G)*, or *equal to (E)* the second. Write a phrase from the word bank to compare the numbers.

a.	b. 2 tens 3 tens
20 ______________ 18	2 tens ______________ 3 tens
c. 24 15	d. 26 32
24 ______________ 15	26 ______________ 32

2. Write a phrase from the word bank to compare the numbers.

36 ______________________ 3 tens 6 ones

1 ten 8 ones ______________________ 3 tens 1 one

38 ______________________ 26

1 ten 7 ones ______________________ 27

15 ______________________ 1 ten 2 ones

30 ______________________ 28

29 ______________________ 32

3. Put the following numbers in order from *least* to *greatest*. Cross off each number after it has been used.

9	40	32	13	23

4. Put the following numbers in order from *greatest* to *least*. Cross off each number after it has been used.

9	40	32	13	23

5. Use the digits 8, 3, 2, and 7 to make 4 different two-digit numbers less than 40. Write them in order from *greatest* to *least*.

8 3 2 7

Examples: 32, 27,...

EUREKA MATH™

Name ______________________________ Date ______________

Word Bank

is greater than
is less than
is equal to

1. Draw the numbers using quick tens and circles. Use the phrases from the word bank to complete the sentence frames to compare the numbers. The first one has been done for you.

a. 20 30 20 ___is less than___ 30	b. 14 22 14 ______________ 22
c. 15 1 ten 5 ones 15 ______________ 1 ten 5 ones	d. 39 29 39 ______________ 29
e. 31 13 31 ______________ 13	f. 23 33 23 ______________ 33

2. Circle the numbers that are *greater* than 28.

32 29 2 tens 8 ones 4 tens 18

3. Circle the numbers that are *less* than 31.

29 3 tens 6 ones 3 tens 13 3 tens 9 ones

4. Write the numbers in order from *least* to *greatest*.

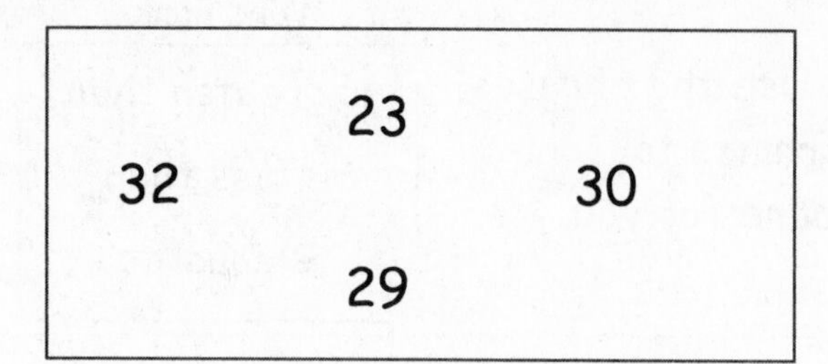

______ ______ ______ ______

Where would the number 27 go in this order? Use words or rewrite the numbers to explain.

5. Write the numbers in order from *greatest* to *least*.

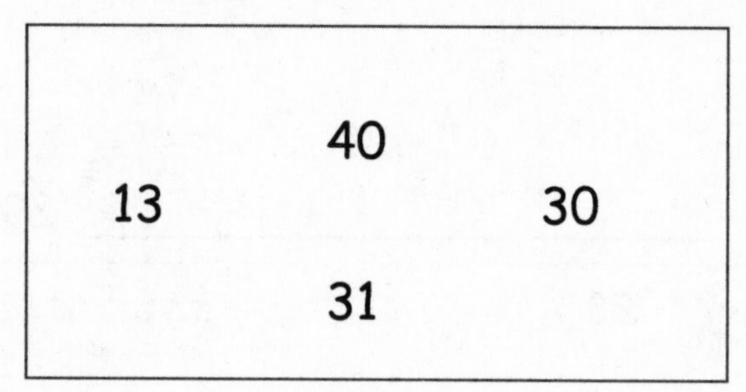

______ ______ ______ ______

Where would the number 23 go in this order? Use words or rewrite the numbers to explain.

6. Use the digits 9, 4, 3, and 2 to make 4 different two-digit numbers less than 40. Write them in order from *least* to *greatest*.

9 3 4 2

Examples: 34, 29,...

Name ____________________ Date ____________

1. Circle the alligator that is eating the *greater* number.

a.	b.	c.	d.
40 > < 20	10 > < 30	18 > < 14	19 > < 36

2. Write the numbers in the blanks so that the alligator is eating the *greater* number. With a partner, compare the numbers out loud, using *is greater than, is less than,* or *is equal to.* Remember to start with the number on the left.

a. 24 4	b. 38 36	c. 15 14
____ > ____	____ < ____	____ < ____
d. 20 2	e. 36 35	f. 20 19
____ > ____	____ < ____	____ > ____
g. 31 13	h. 23 32	i. 21 12
____ > ____	____ < ____	____ < ____

3. If the alligator is eating the *greater* number, circle it. If not, redraw the alligator.

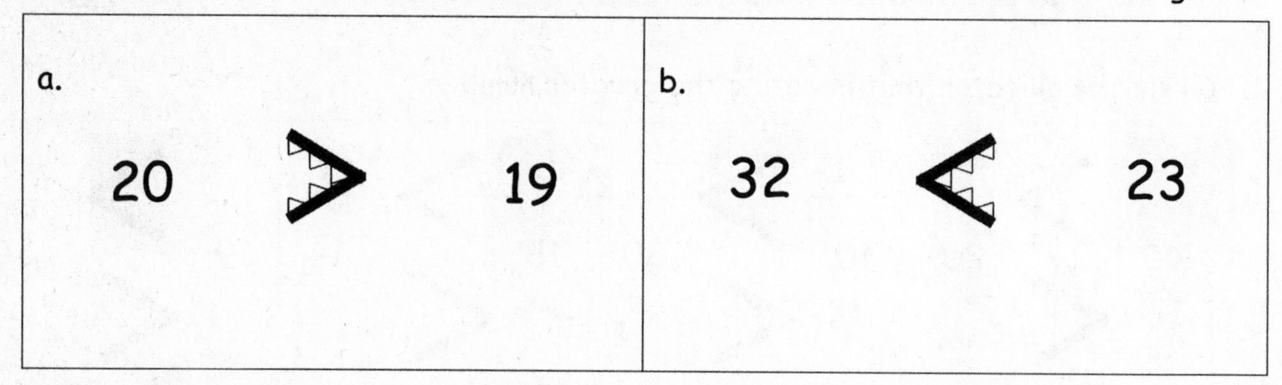

4. Complete the charts so that the alligator is eating a *greater* number.

	tens	ones		tens	ones
a.	1	2	>	1	
b.	2	7	>	2	
c.	2	5	>		5
d.		8	<	3	8
e.	2	1	>	2	
f.	2	4	<		4
g.	1	8	>		5
h.	2	1	>		9
i.		7	<	2	1
j.	1	4	>		4

Name ______________________________ Date ______________

1. Write the numbers in the blanks so that the alligator is eating the greater number. Read the number sentence, using *is greater than, is less than,* or *is equal to.* Remember to start with the number on the left.

a. 10 20	b. 15 17	c. 24 22
____ $>$ ____	____ $<$ ____	____ $>$ ____
d. 29 30	e. 39 38	f. 39 40
____ $>$ ____	____ $<$ ____	____ $<$ ____

2. Complete the charts so that the alligator is eating a *greater* number.

	tens	ones		tens	ones
a.	1	8	$>$	1	
b.	2	4	$<$		3
c.			$>$		
d.	2	3	$>$	2	
e.			$<$		
f.	1	7	$>$		7

Compare each set of numbers by matching to the correct alligator or phrase to make a true number sentence. Check your work by reading the sentence from left to right.

3.

16	17

31	23

35	25

12	21

22	32

29	30

39	40

is *less* than

is *greater* than

Name ______________________________ Date ____________

1. Use the symbols to compare the numbers. Fill in the blank with <, >, or = to make a true number sentence. Read the number sentences from left to right.

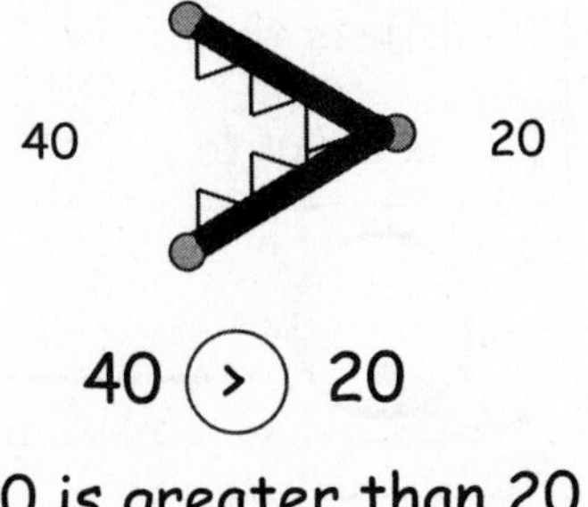

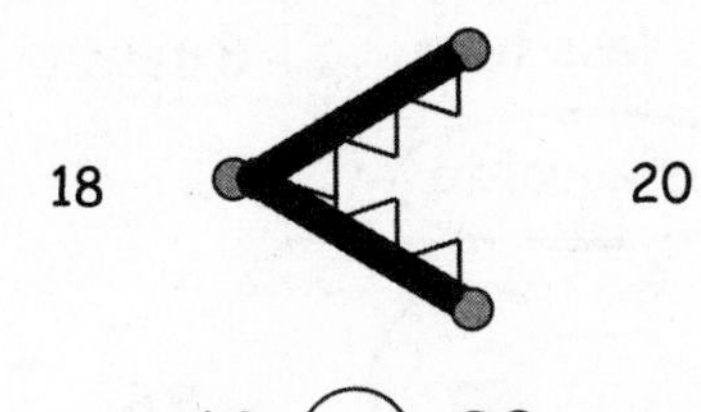

40 (>) 20

40 is greater than 20.

18 (<) 20

18 is less than 20.

a.	b.	c.
27 ◯ 24	31 ◯ 28	10 ◯ 13
d.	e.	f.
13 ◯ 15	31 ◯ 29	38 ◯ 18
g.	h.	i.
27 ◯ 17	32 ◯ 21	12 ◯ 21

2. Circle the correct words to make the sentence true. Use >, <, or = and numbers to write a true number sentence. The first one is done for you.

a. 36 [is greater than / is less than / **is equal to** (circled)] 3 tens 6 ones 36 (=) 36	b. 1 ten 4 ones [is greater than / is less than / is equal to] 17 ____ ( ) ____
c. 2 tens 4 ones [is greater than / is less than / is equal to] 34 ____ ( ) ____	d. 20 [is greater than / is less than / is equal to] 2 tens 0 ones ____ ( ) ____
e. 31 [is greater than / is less than / is equal to] 13 ____ ( ) ____	f. 12 [is greater than / is less than / is equal to] 21 ____ ( ) ____
g. 17 [is greater than / is less than / is equal to] 3 ones 1 ten ____ ( ) ____	h. 30 [is greater than / is less than / is equal to] 0 tens 30 ones ____ ( ) ____

Name ______________________________ Date ______________

Use the symbols to compare the numbers. Fill in the blank with <, >, or = to make a true number sentence. Complete the number sentence with a phrase from the word bank.

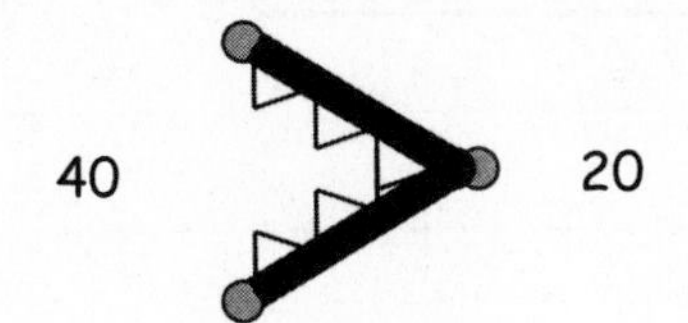

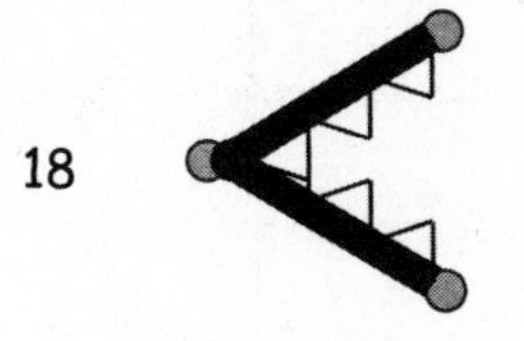

20

Word Bank
is greater than
is less than
is equal to

40 (>) 20
40 is greater than 20.

18 (<) 20
18 is less than 20.

a. 17 ◯ 13

17 ______________ 13

b. 23 ◯ 33

23 ______________ 33

c. 36 ◯ 36

36 ______________ 36

d. 25 ◯ 32

25 ______________ 32

e. 38 ◯ 28

38 ______________ 28

f. 32 ◯ 23

32 ______________ 23

g. 1 ten 5 ones ◯ 14

1 ten 5 ones ________ 14

h. 3 tens ◯ 30

3 tens ____________ 30

i. 29 ◯ 2 tens 7 ones

29 ________ 2 tens 7 ones

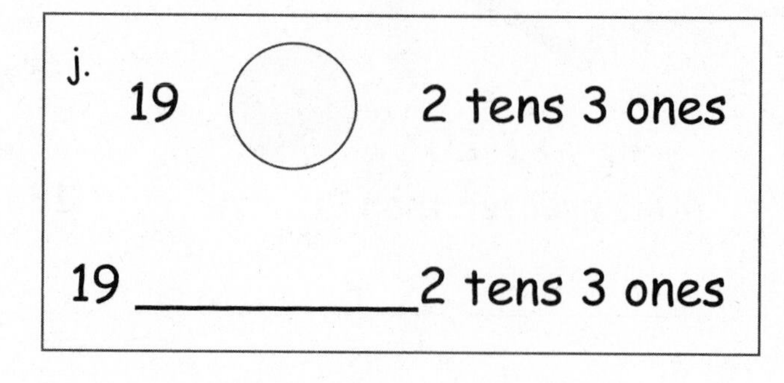

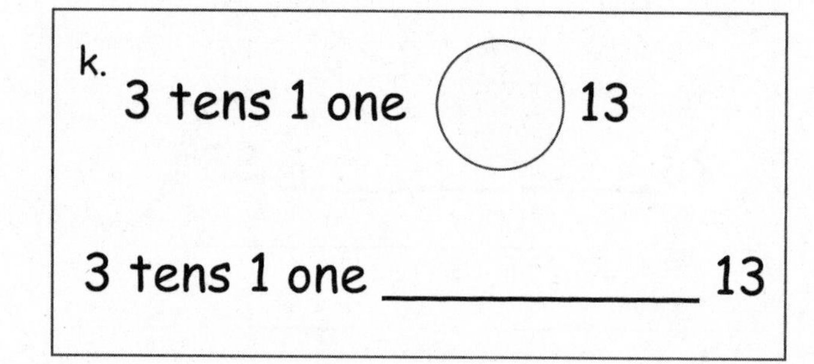

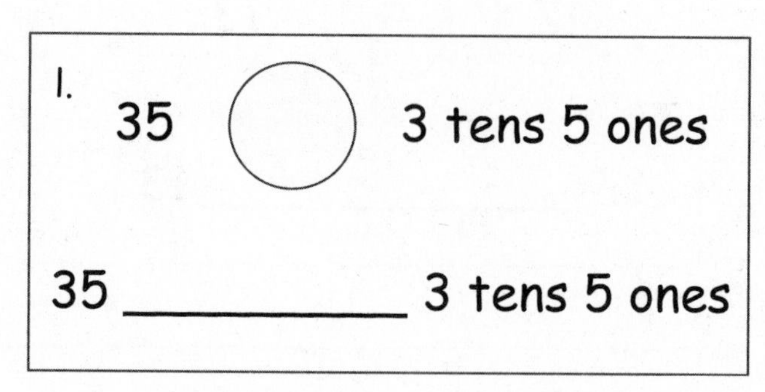

m. 2 tens 3 ones ◯ 32

2 tens 3 ones _________ 32

n. 3 tens ◯ 36

3 tens _____________36

o. 29 ◯ 3 tens 9 ones

29 ________ 3 tens 9 ones

p. 4 tens ◯ 39

4 tens _____________ 39

Name ______________________________ Date ______________

Complete the number bonds and number sentences to match the picture. The first one is done for you.

<table>
<tr>
<td>

1.

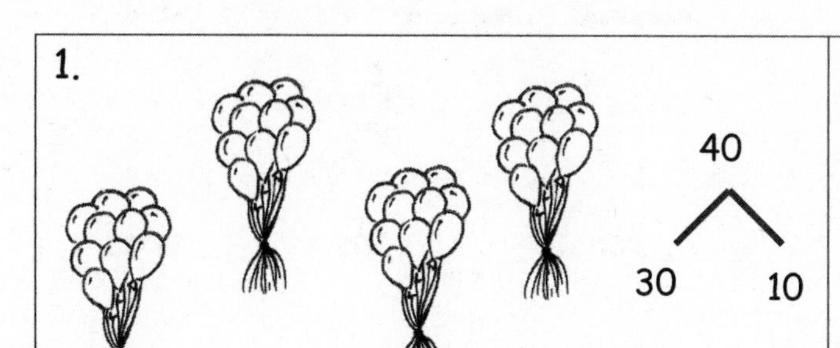

3 tens + 1 ten = 4 tens

30 + 10 = 40

</td>
<td>

2.

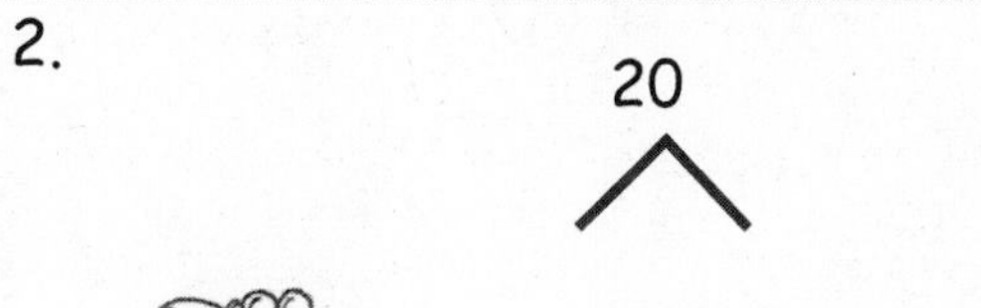

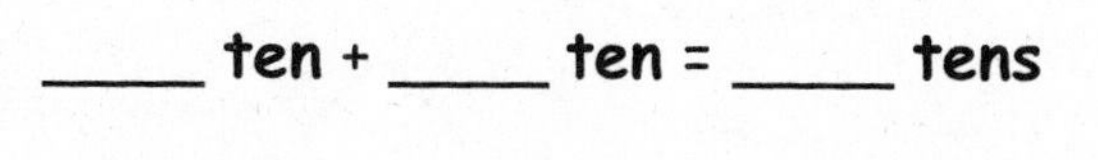

____

</td>
</tr>
<tr>
<td>

3.

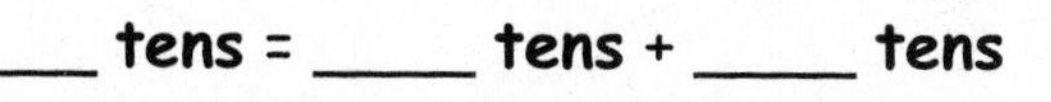

____

</td>
<td>

4.

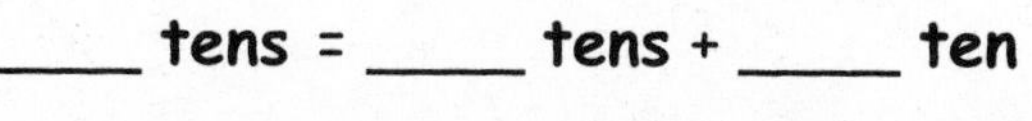

</td>
</tr>
</table>

5.

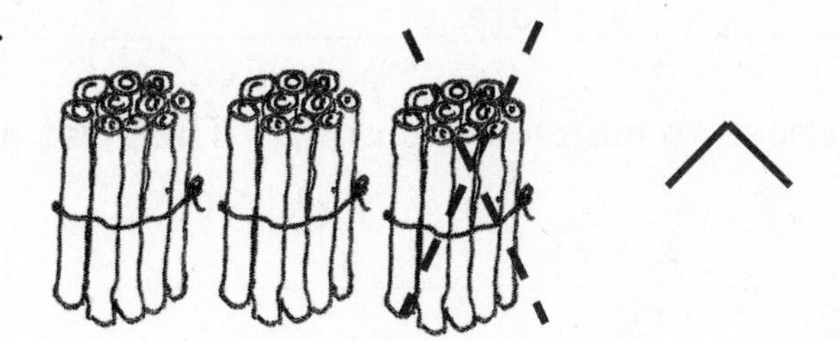

_____ tens - _____ ten = _____ tens

____

6.

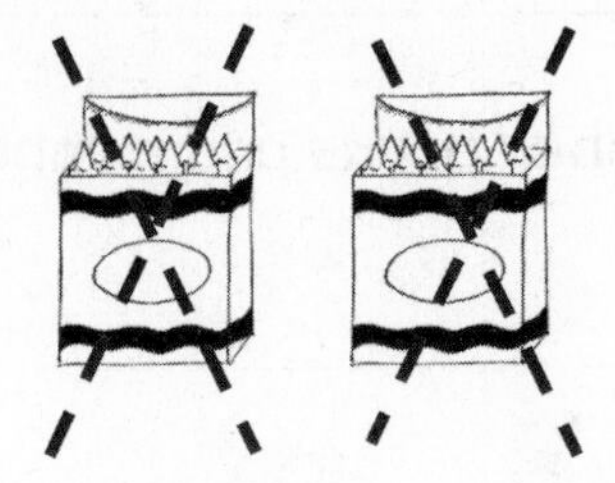

_____ tens - _____ tens = _____ tens

____

7.

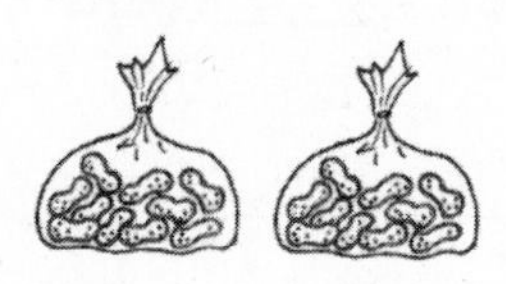

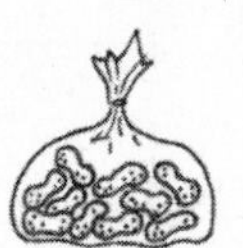

_____ tens + _____ ten = _____ tens

____

8.

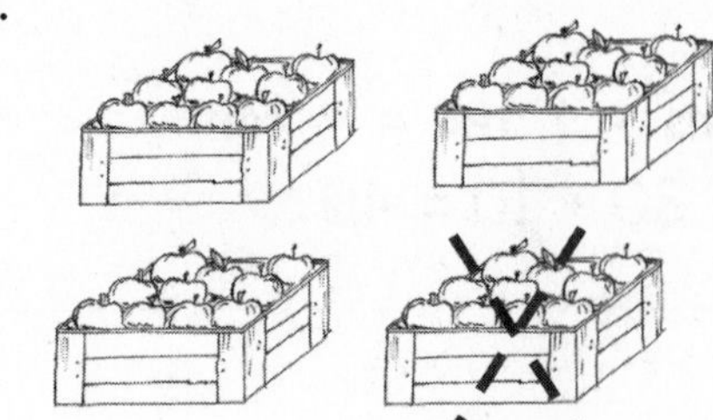

_____ tens - _____ ten = _____ tens

_____+_________________________

9.

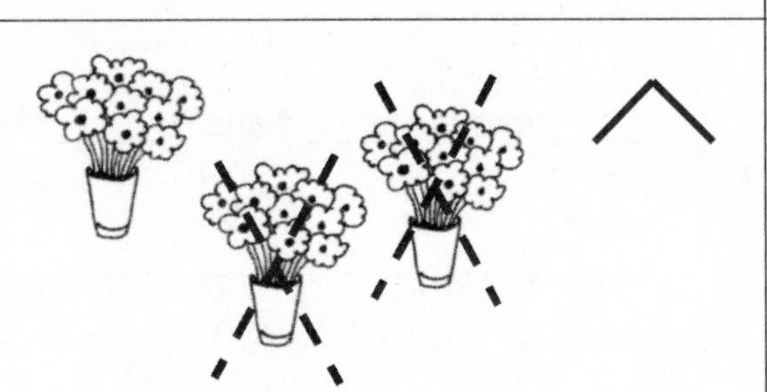

_____ tens - _____ tens = _____ ten

____

10.

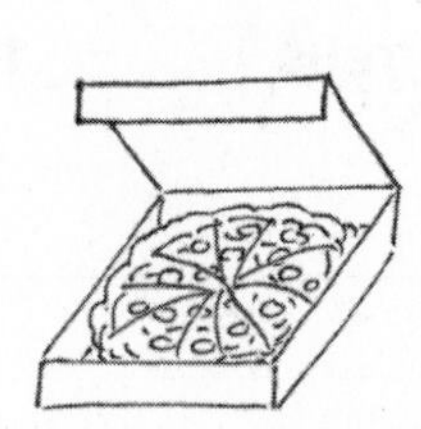

_____ ten - _____ tens = _____ ten

____

11. Fill in the missing numbers. Match the related addition and subtraction facts.

a. 4 tens - 2 tens = ______ 2 tens + 1 ten = 3 tens

b. 40 - 30 = ______ 30 + 10 = 40

c. 30 - 20 = ______ 20 + 20 = 40

12. Fill in the missing numbers.

a. 20 + 20 = ______

b. 30 - 20 = ______

c. 10 + ______ = 40

d. 20 - ______ = 0

e. 40 - ______ = 10

f. _____ + _____ = 30

G1-M4-SE-1.3.1-1.2016

Name ______________________________ Date ______________

Draw a number bond, and complete the number sentences to match the pictures.

1.

30

20 10

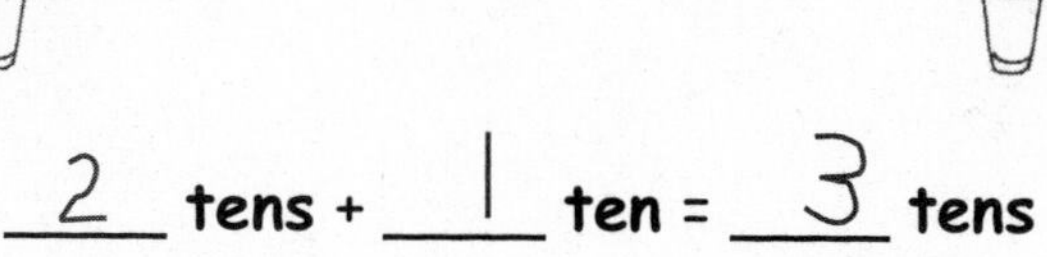

__2__ tens + __1__ ten = __3__ tens

20 + 10 = 30

2.

____ tens = ____ ten + ____ tens

____

3.

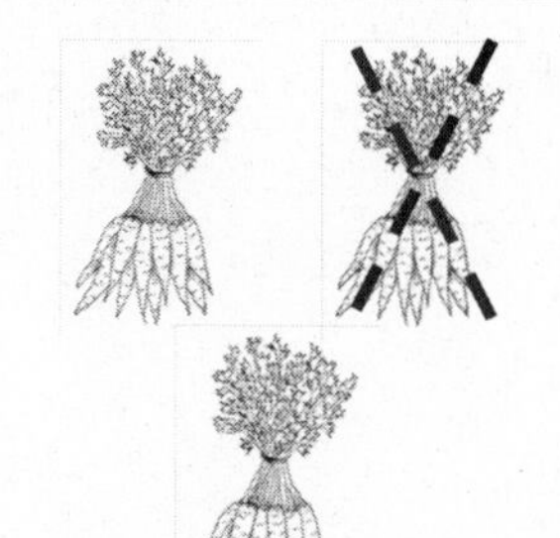

____ tens - ____ ten = ____ tens

____

4.

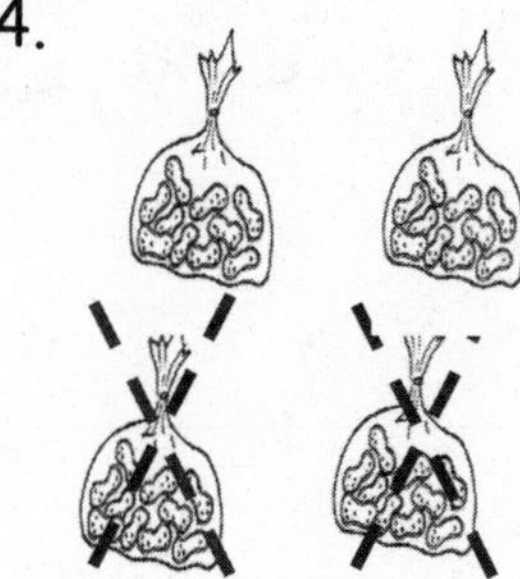

____ tens - ____ tens = ____ tens

____

5.

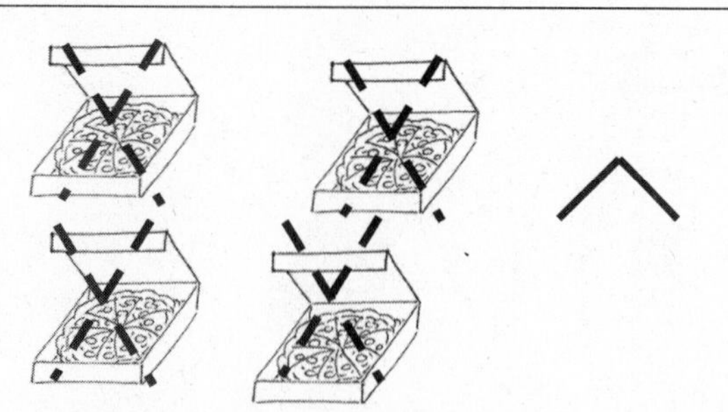

____ tens - ____ tens = ____ tens

____

6.

____ tens + ____ tens = ____ tens

____

Draw quick tens and a number bond to help you solve the number sentences.

7. $10 + 20 =$ ______	8. $30 - 10 =$ ______
9. $20 - 10 =$ ______	10. $30 + 10 =$ ______

Add or subtract.

11. 2 tens + 1 ten = ______

12. $20 + 20 =$ ______

13. $40 - 10 =$ ______

14. ______ $= 20 + 10$

15. 3 tens - 2 tens = ______

16. $20 - 10 =$ ______

17. $10 - 10 =$ ______

18. ______ $= 30 + 10$

19. $40 - 30 =$ ______

This page intentionally left blank

_____ ○ _____ ○ _____

_____ tens ○ _____ tens ○ _____ tens

_____ ○ _____ ○ _____

number bond/number sentence set

This page intentionally left blank

Name ______________________________ Date ______________

Fill in the missing numbers to match the picture. Write the matching number bond.

1.

32

12 20

12 + 20 = ______

2.

15 + ______ = ______

3.

______ + ______ = ______

4.

______ + ______ = ______

Draw using quick tens and ones. Complete the number bond, and write the sum in the place value chart and the number sentence.

5.

19 + 10 = _____

tens	ones

6.

20 + 14 = _____

tens	ones

Use arrow notation to solve.

7. 13 $\xrightarrow{+10}$ ____	8. 19 $\xrightarrow{+\ }$ 39
9. ____ $\xrightarrow{+10}$ 26	10. ____ $\xrightarrow{+20}$ 38

Use the dimes and pennies to complete the place value charts and the number sentences.

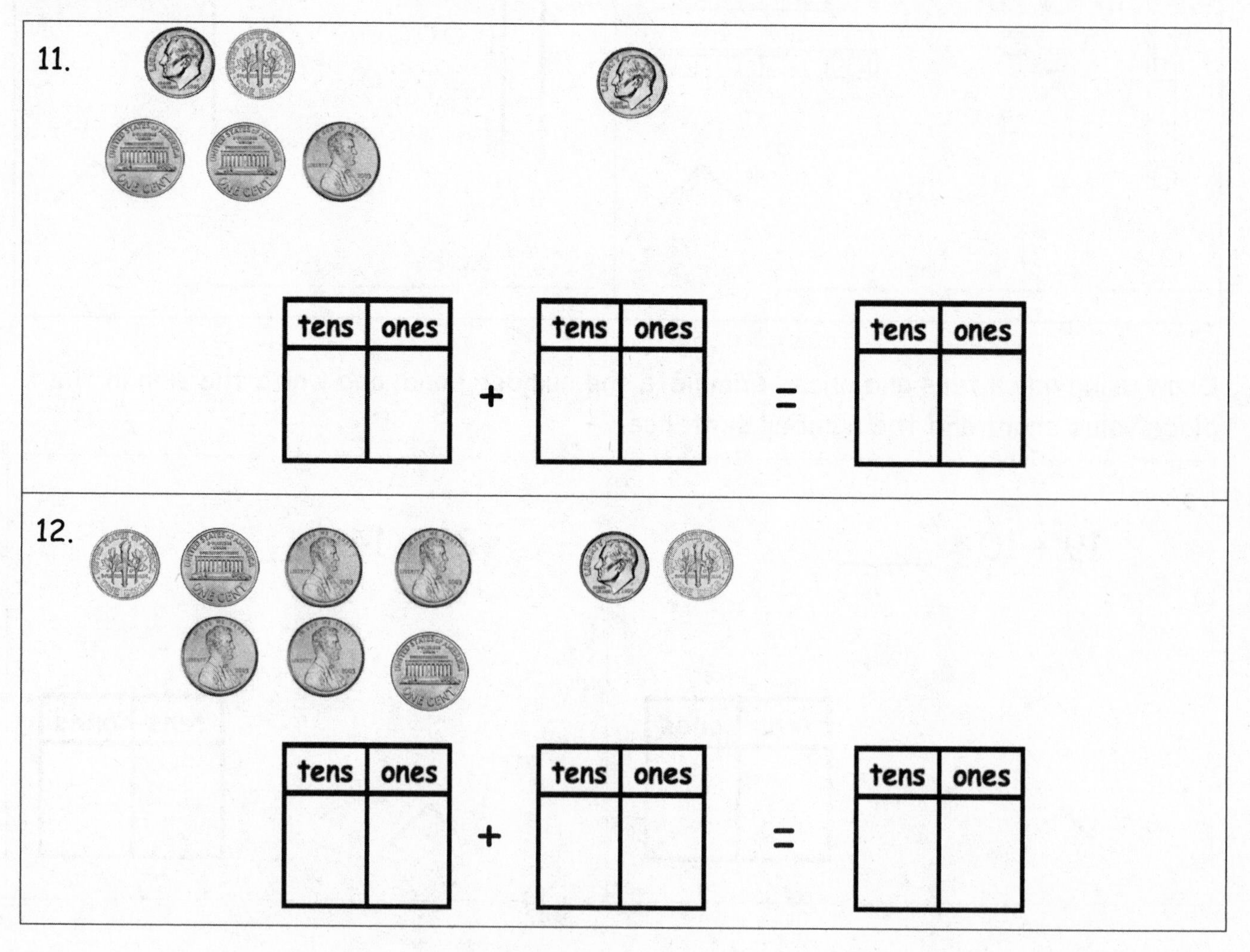

Name ______________________________ Date ________________

Fill in the missing numbers to match the picture. Complete the number bond to match.

1.

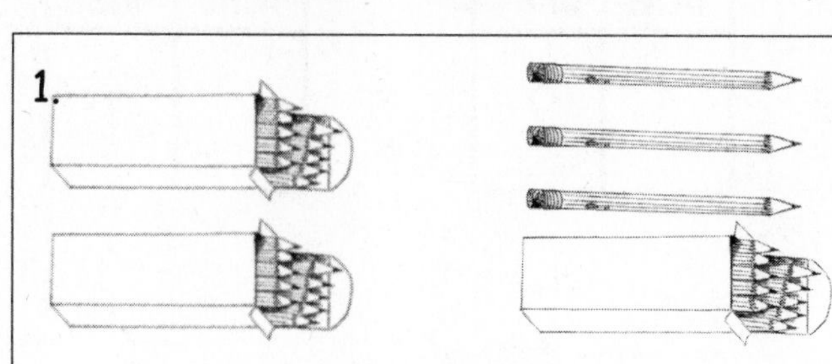

$20 + 13 =$ ____

2.

$17 +$ ____ $=$ ____

3.

____ + ____ = ____

4.

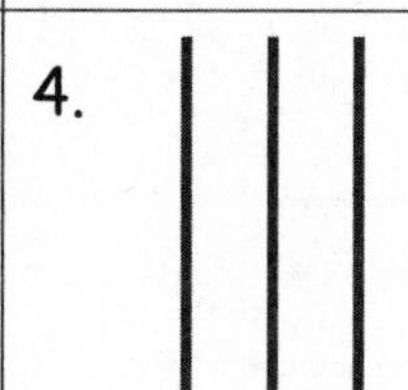

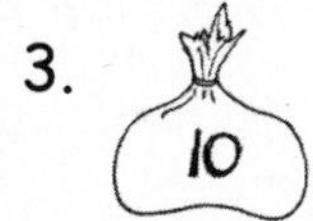

____ + ____ = ____

Draw using quick tens and ones. Complete the number bond and the number sentence.

5.

tens	ones
1	7

\+

tens	ones
1	0

_____ + _____ = _____

6.

tens	ones
1	9

\+

tens	ones

_____ + _____ = 39

Use arrow notation to solve.

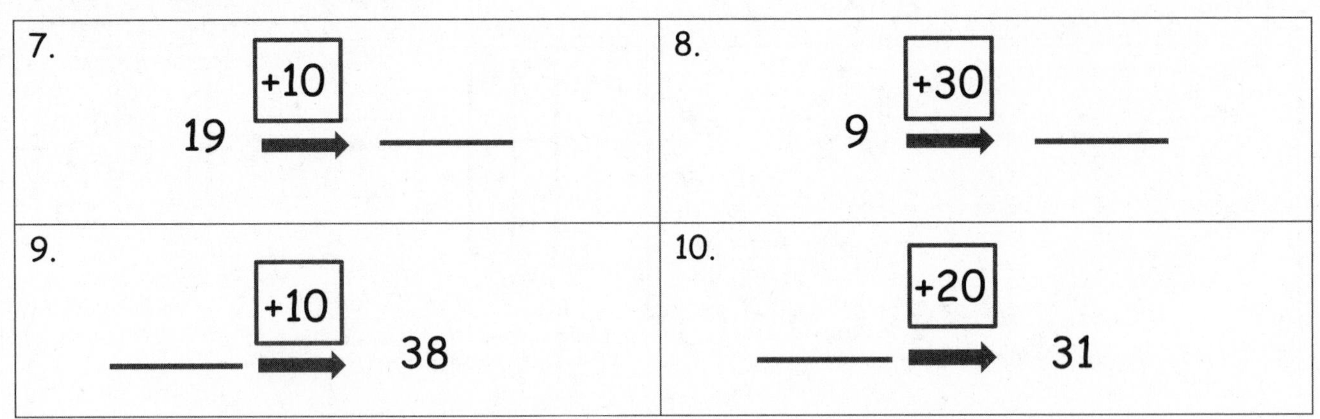

Use the dimes and pennies to complete the place value charts.

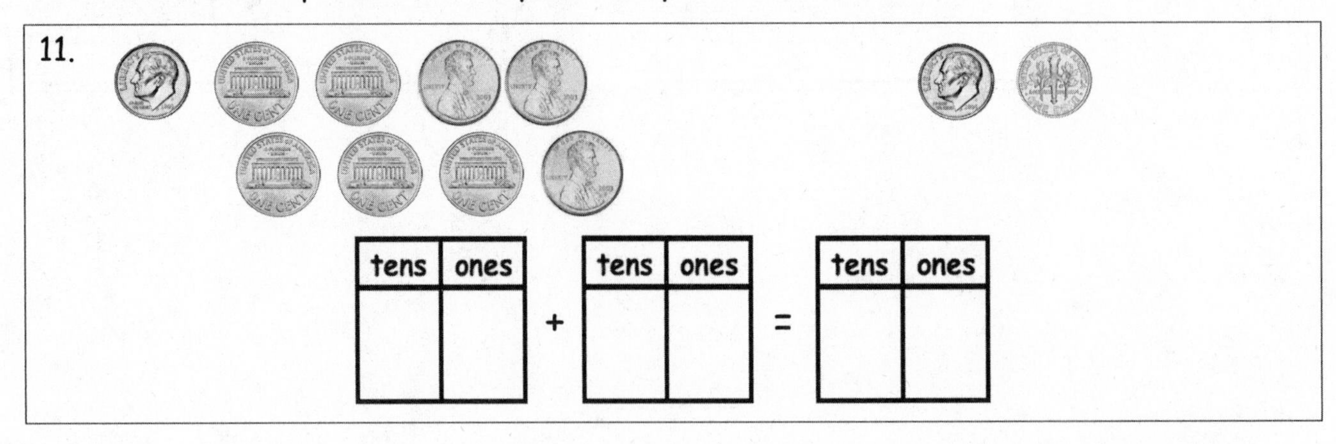

EUREKA MATH™

Name ______________________________ Date ______________

Use the pictures to complete the place value chart and number sentence. For Problems 5 and 6, make a quick ten drawing to help you solve.

1.

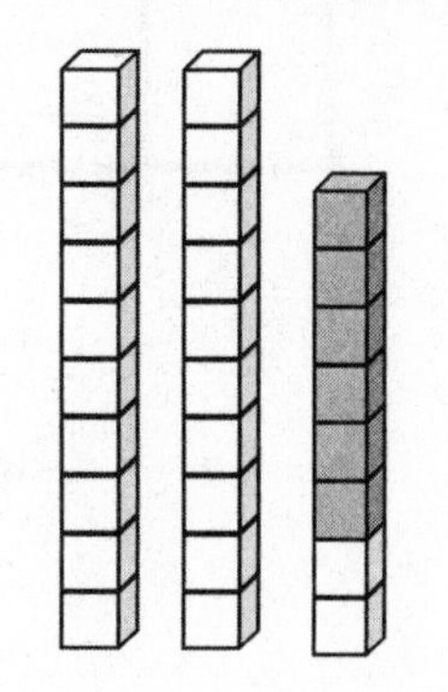

tens	ones

22 + 6 = ______

2.

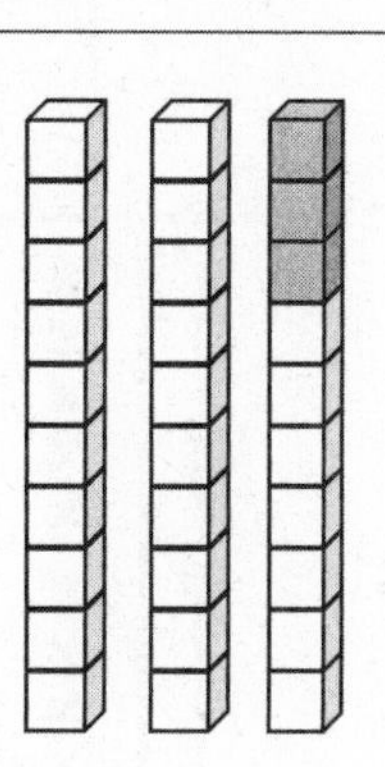

tens	ones

______ + 3 = ______

3.

tens	ones

12 + ______ = ______

4.

tens	ones

______ + ______ = ______

5.

tens	ones

24 + 6 = ______

6.

tens	ones

24 + 3 = ______

Draw quick tens, ones, *and* number bonds to solve. Complete the place value chart.

7. 21 + 9 = ______

tens	ones

8. 21 + 7 = ______

tens	ones

9. 13 + 7 = ______

tens	ones

10. 26 + 4 = ______

tens	ones

11. 32 + 3 = ______

tens	ones

12. 38 + 2 = ______

tens	ones

Name ______________________________ Date ______________

Use quick tens and ones to complete the place value chart and number sentence.

1.

tens	ones

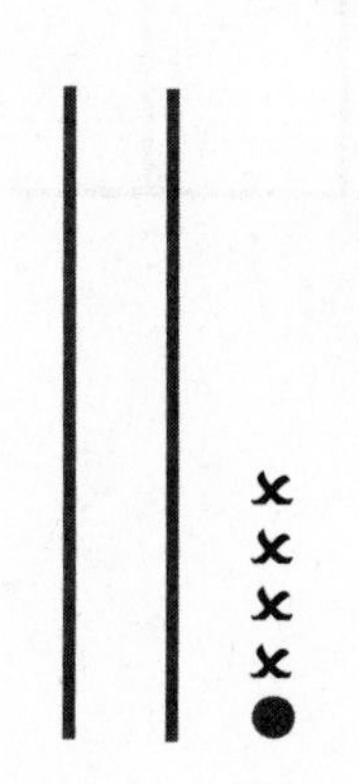

21 + 4 = ______

2.

tens	ones

21 + 8 = ______

3.

tens	ones

25 + 4 = ______

4.

tens	ones

25 + 5 = ______

5.

tens	ones

33 + 3 = ______

6.

tens	ones

33 + 7 = ______

Draw quick tens, ones, and number bonds to solve. Complete the place value chart.

7. 26 + 2 = ______

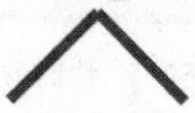

tens	ones

8. 36 + 3 = ______

tens	ones

9. 26 + 4 = ______

tens	ones

10. 24 + 6 = ______

tens	ones

11. Solve. You may draw quick tens and ones or number bonds to help.

a. 22 + 7 = ______ b. 22 + 8 = ______ c. 32 + 8 = ______

Name ________________________________ Date ________________

Use the pictures or draw quick tens and ones. Complete the number sentence and place value chart.

1. 18 + 1 = ______

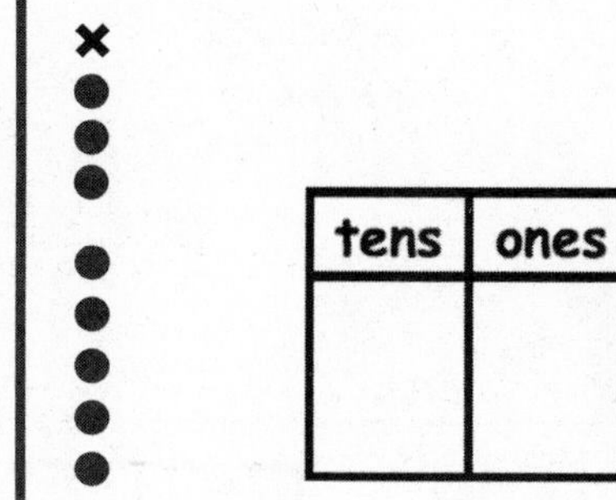

2. 18 + 2 = ______

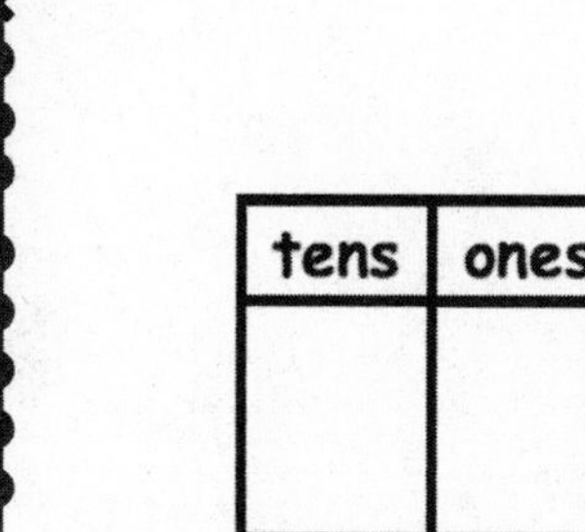

3. 18 + 5 = ______

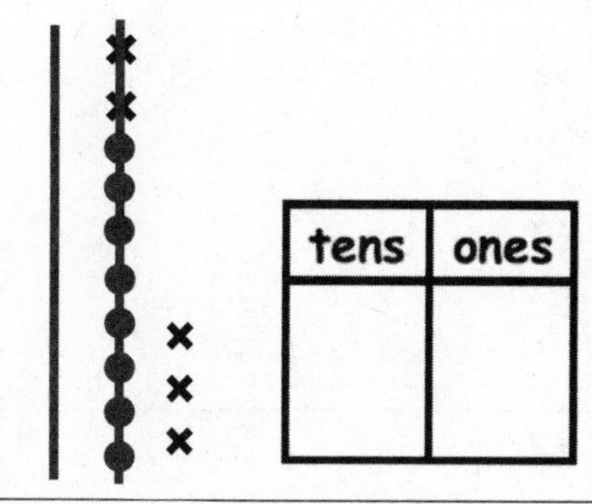

4. 29 + 1 = ______

tens	ones

5. 29 + 3 = ______

tens	ones

6. 29 + 6 = ______

tens	ones

7. 16 + 4 = ______

tens	ones

8. 16 + 6 = ______

tens	ones

9. 26 + 6 = ______

tens	ones

Make a number bond to solve. Show your thinking with number sentences or the arrow way. Complete the place value chart.

10. $17 + 2 =$ ______

tens	ones

11. $17 + 5 =$ ______

tens	ones

12. $25 + 4 =$ ______

tens	ones

13. $25 + 6 =$ ______

tens	ones

14. $34 + 4 =$ ______

tens	ones

15. $34 + 8 =$ ______

tens	ones

Name ______________________ Date ______________

Use the pictures or draw quick tens and ones. Complete the number sentence and place value chart.

1. 15 + 3 = ______

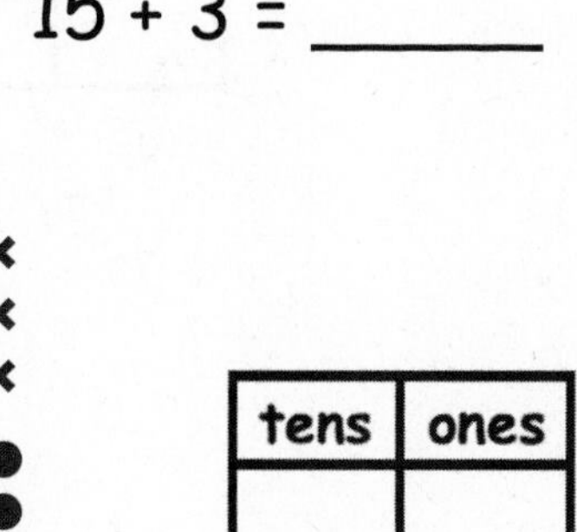

tens	ones

2. 15 + 5 = ______

tens	ones

3. 15 + 6 = ______

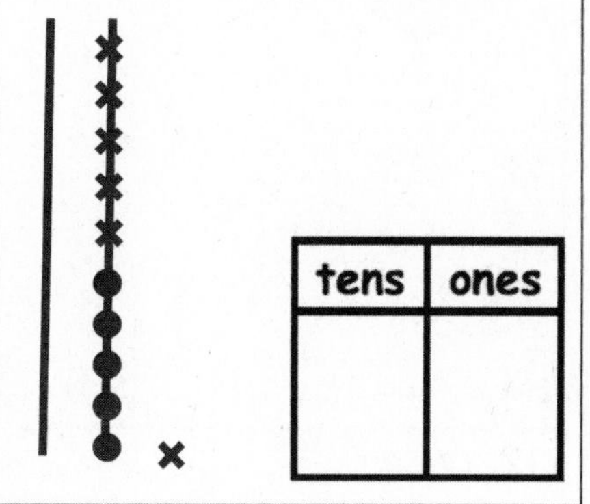

tens	ones

4. 28 + 2 = ______

tens	ones

5. 28 + 4 = ______

tens	ones

6. 28 + 7 = ______

tens	ones

7. 17 + 3 = ______

tens	ones

8. 17 + 7 = ______

tens	ones

9. 27 + 7 = ______

tens	ones

Make a number bond to solve. Show your thinking with number sentences or the arrow way. Complete the place value chart.

10. 13 + 6 = ______

tens	ones

11. 13 + 7 = ______

tens	ones

12. 25 + 5 = ______

tens	ones

13. 25 + 8 = ______

tens	ones

14. 24 + 8 = ______

tens	ones

15. 23 + 9 = ______

tens	ones

Name ______________________________ Date ________________

Solve the problems.

1.	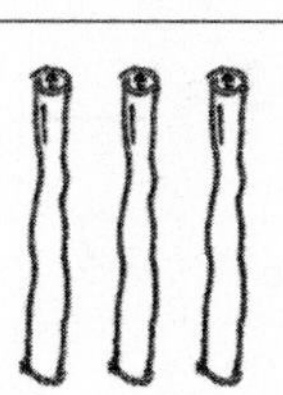	5 + 3 = _____
2.	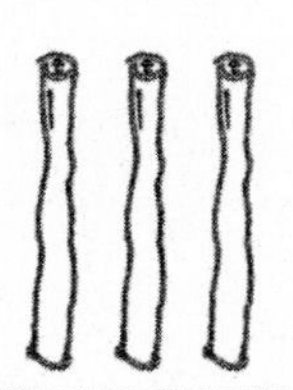	15 + 3 = _____
3.	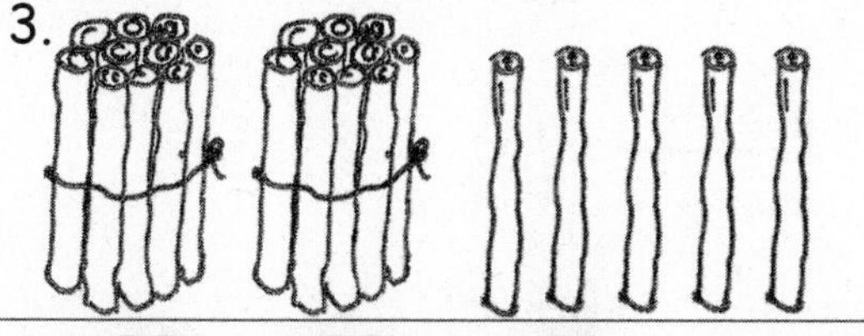 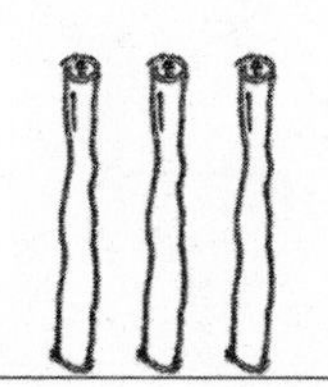	25 + 3 = _____
4.	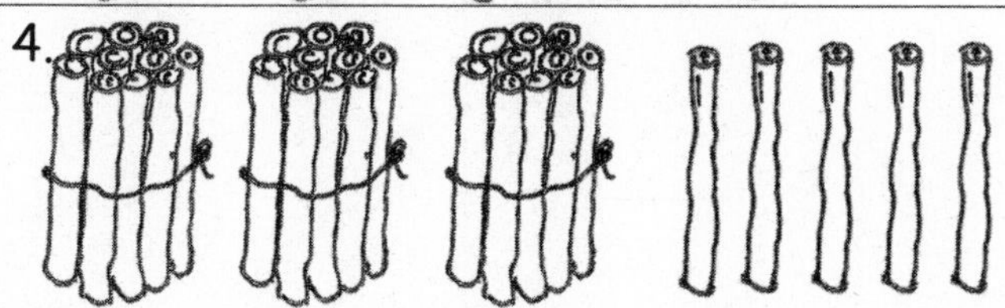 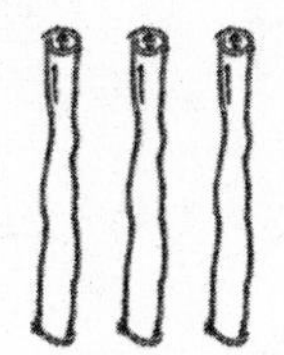	35 + 3 = _____
5.	 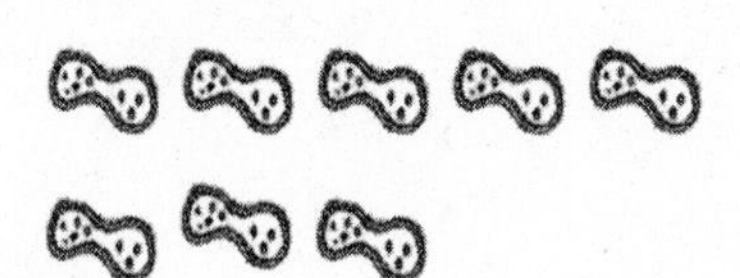	8 + 4 = _____
6.	 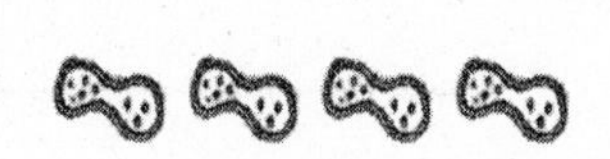	18 + 4 = _____
7.		28 + 4 = _____

8. Solve the problems.

a. 6 + 2 = _____	b. 16 + 2 = _____	c. 26 + 2 = _____	d. 36 + 2 = _____
e. 6 + 4 = _____	f. 16 + 4 = _____	g. 26 + 4 = _____	h. 36 + 4 = _____
i. 9 + 2 = _____	j. 19 + 2 = _____	k. 29 + 2 = _____	
l. 8 + 6 = _____	m. 18 + 6 = _____	n. 28 + 6 = _____	

Solve the problems. Show the 1-digit addition sentence that helped you solve.

9. 23 + 6 = _____ ________________________________

10. 27 + 6 = _____ ________________________________

Name ______________________________ Date ______________

Solve the problems.

1. 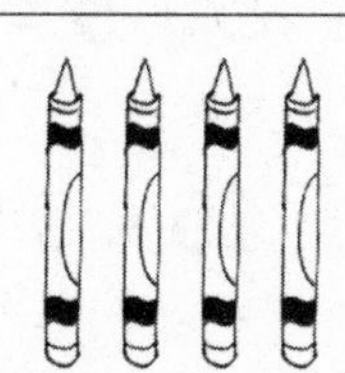

$5 + 4 =$ ____

2. 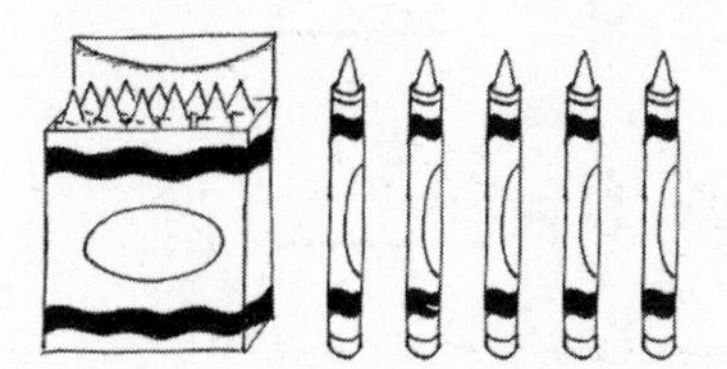 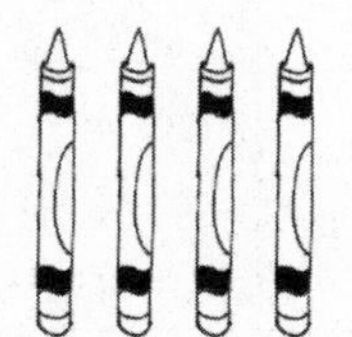

$15 + 4 =$ ____

3. 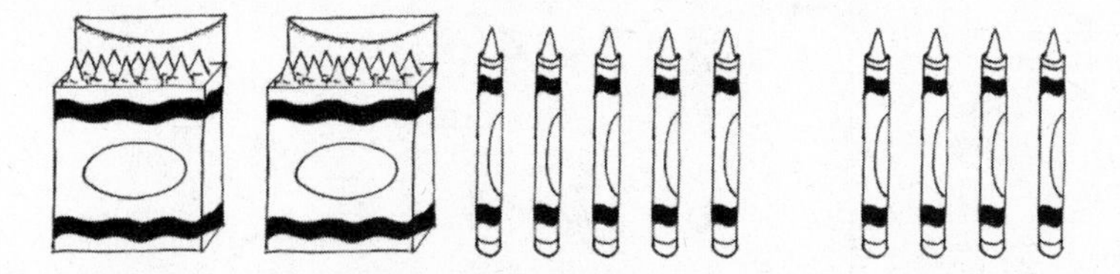

$25 + 4 =$ ____

4. 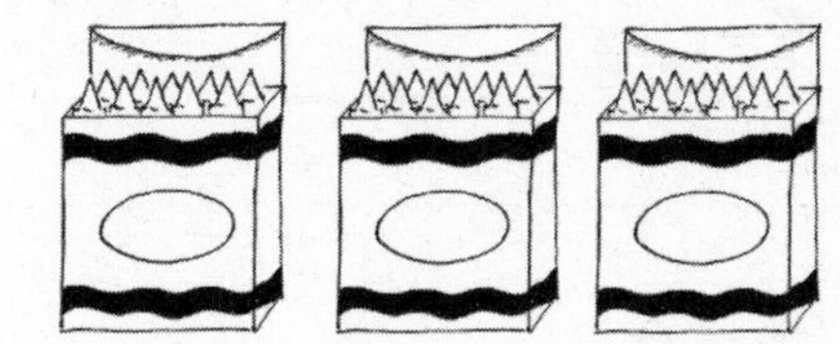 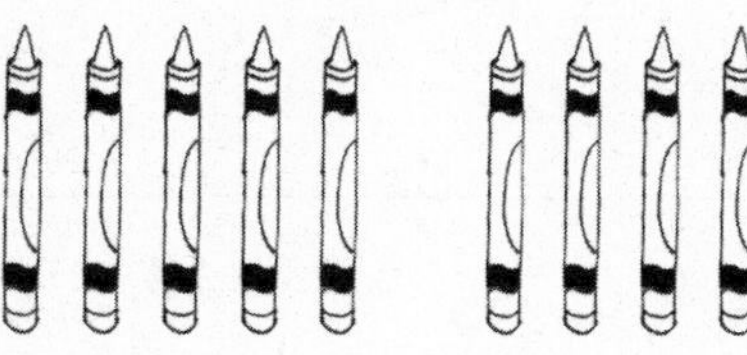

$35 + 4 =$ ____

5. 

$8 + 4 =$ ____

6.  

$18 + 4 =$ ____

7. 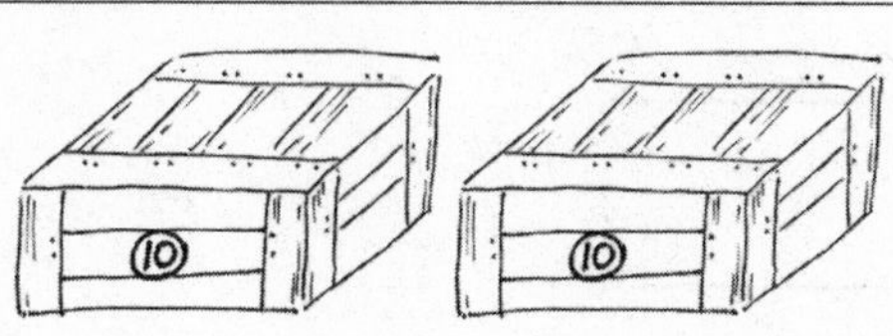 

$28 + 4 =$ ____

Use the first number sentence in each set to help you solve the other problems.

8.	9.
a. 5 + 2 = ____	a. 5 + 5 = ____
b. 15 + 2 = ____	b. 15 + 5 = ____
c. 25 + 2 = ____	c. 25 + 5 = ____
d. 35 + 2 = ____	d. 35 + 5 = ____

10.	11.
a. 2 + 7 = ____	a. 7 + 4 = ____
b. 12 + 7 = ____	b. 17 + 4 = ____
c. 22 + 7 = ____	c. 27 + 4 = ____

12.	13.
a. 8 + 7 = ____	a. 3 + 9 = ____
b. 18 + 7 = ____	b. 13 + 9 = ____
c. 28 + 7 = ____	c. 23 + 9 = ____

Solve the problems. Show the 1-digit addition sentence that helped you solve.

14. 24 + 5 = ____ ________________________

15. 24 + 7 = ____ ________________________

Name ______________________ Date __________

Draw quick tens and ones to help you solve the addition problems.

1. 16 + 3 = ____	2. 17 + 3 = ____
3. 18 + 20 = ____	4. 31 + 8 = ____
5. 3 + 14 = ____	6. 6 + 30 = ____
7. 23 + 7 = ____	8. 17 + 3 = ____

With a partner, try more problems using quick ten drawings, number bonds, or the arrow way.

9. 32 + 7 = ______

10. 13 + 20 = ______

11. 6 + 34 = ______

12. 4 + 36 = ______

13. 20 + 18 = ______

14. 14 + 20 = ______

15. Draw dimes and pennies to help you solve the addition problems.

a. 16 + 20 = _____	b. 22 + 7 = _____

Name ______________________________ Date ______________

Draw quick tens and ones to help you solve the addition problems.

1. 17 + 2 = ______	2. 17 + 3 = ______
3. 14 + 3 = ______	4. 24 + 10 = ______

Make a number bond or use the arrow way to solve the addition problems.

5. 6 + 24 = ______	6. 14 + 20 = ______

7. Solve each addition sentence, and match.

a.

22 + 1 = ______

b.

13 + 6 = ______

c.

3 + 26 = ______

d.

37 + 3 = ______

e.

22 + 10 = ______

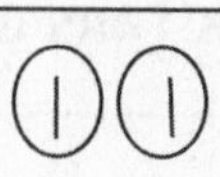

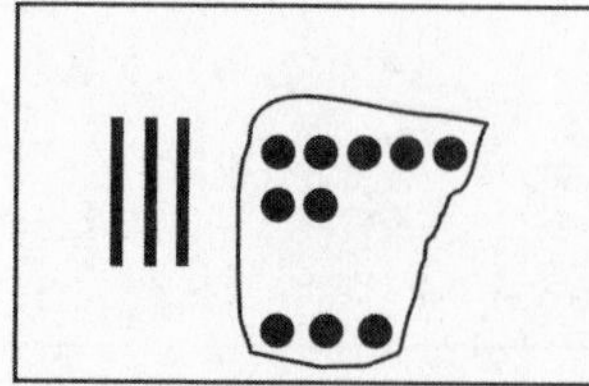

$26 \xrightarrow{+3} 29$

13 + 6

10 3

Name ______________________________ Date ______________

Solve the problems by drawing quick tens and ones or a number bond.

1. 25 + 1 = ____	2. 25 + 10 = ____
3. 15 + 4 = ____	4. 15 + 20 = ____
5. 16 + 7 = ____	6. 26 + 7 = ____
7. 23 + 7 = ____	8. 33 + 7 = ____

9. 16 + 20 = _____	10. 6 + 24 = _____

11. Try more problems with a partner. Use your personal white board to help you solve.

a. 4 + 26

b. 28 + 4

c. 32 + 7

d. 20 + 18

e. 9 + 23

f. 9 + 27

Choose one problem you solved by drawing quick tens, and be ready to discuss.

Choose one problem you solved using the number bond, and be ready to discuss.

EUREKA MATH™

Name ______________________________ Date ______________

Use quick ten drawings or number bonds to make true number sentences.

1. 13 + 20 = _____	2. 23 + 6 = _____
3. 10 + 23 = _____	4. 28 + 6 = _____
5. 26 + 7 = _____	6. 20 + 17 = _____

7. How did you solve Problem 5? Why did you choose to solve it that way?

Solve using quick ten drawings or number bonds.

8. $23 + 9 =$ ______	9. $27 + 7 =$ ______
10. $24 + 10 =$ ______	11. $20 + 18 =$ ______
12. $28 + 9 =$ ______	13. $29 + 9 =$ ______

14. How did you solve Problem 11? Why did you choose to solve it that way?

Name ______________________________ Date ______________

1. Each of the solutions is missing numbers or parts of the drawing. Fix each one so it is accurate and complete.

B+20=2

13 + 8 = 21

a.

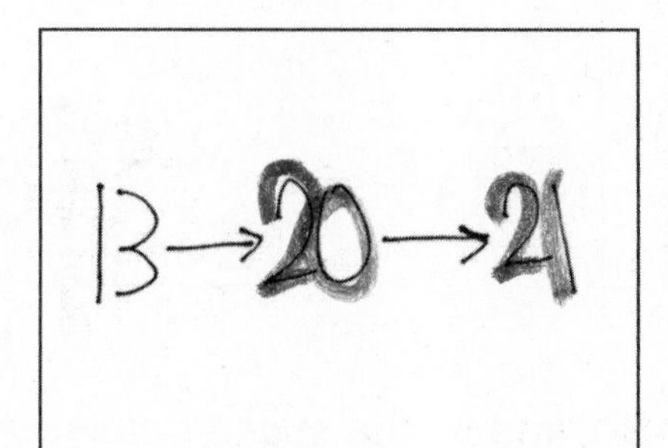

b.

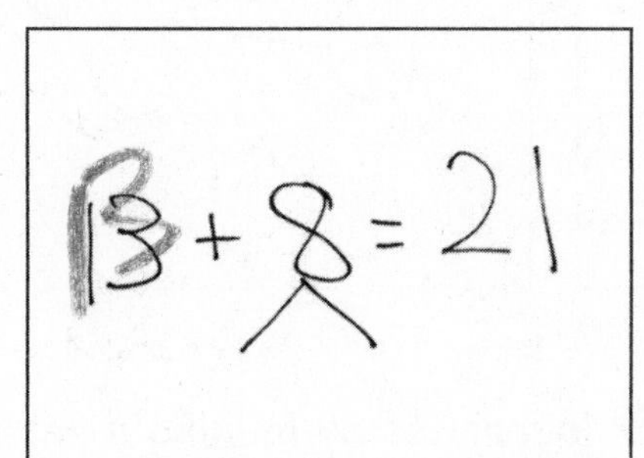

c.

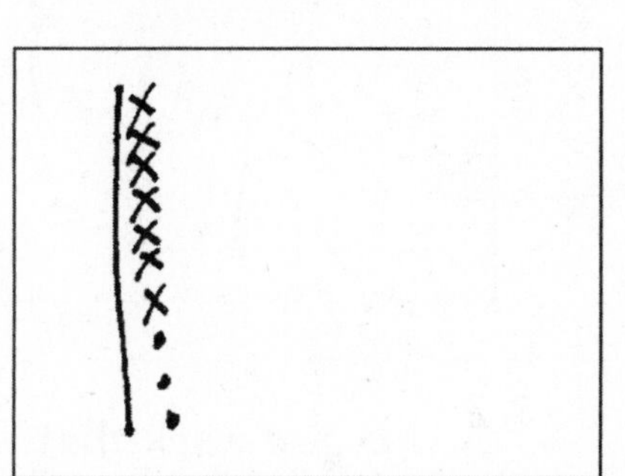

2. Circle the student work that correctly solves the addition problem.

16 + 5

a.

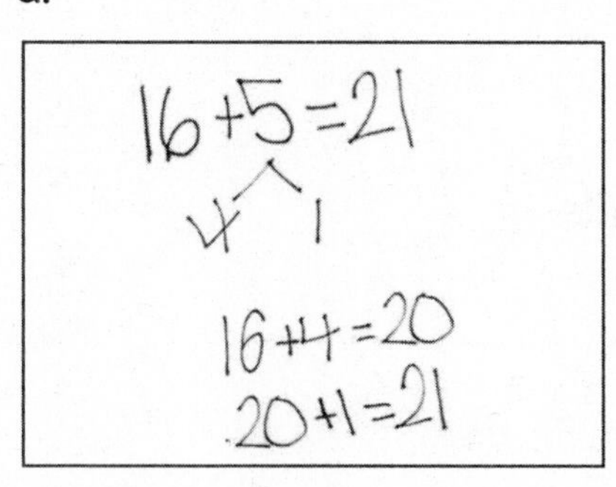

b.

c.

16 +3 → 20 +2 → 22

d. Fix the work that was incorrect by making new work in the space below with the matching number sentence.

3. Circle the student work that correctly solves the addition problem.

$$13 + 20$$

a.

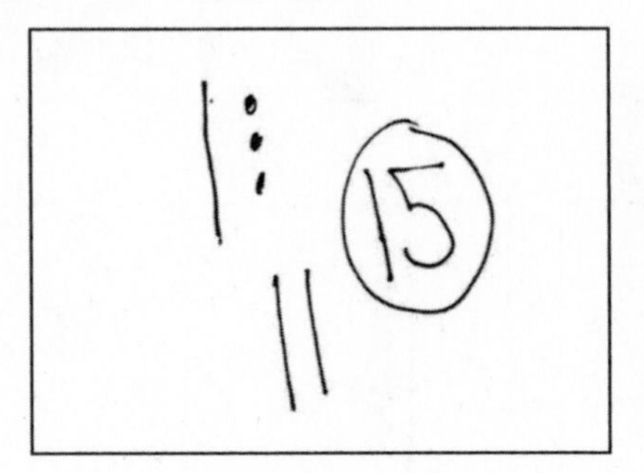

b.

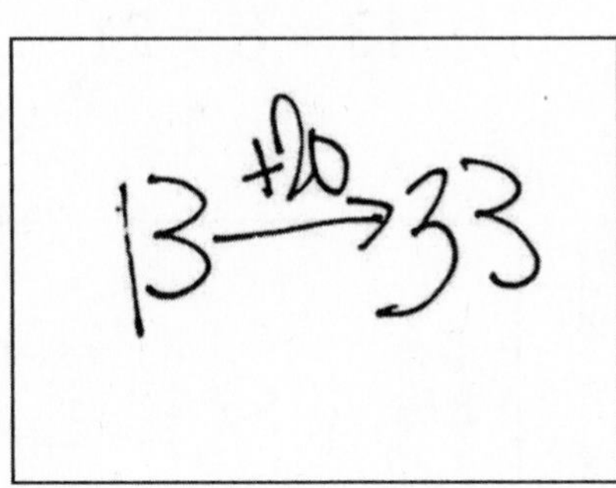

c.

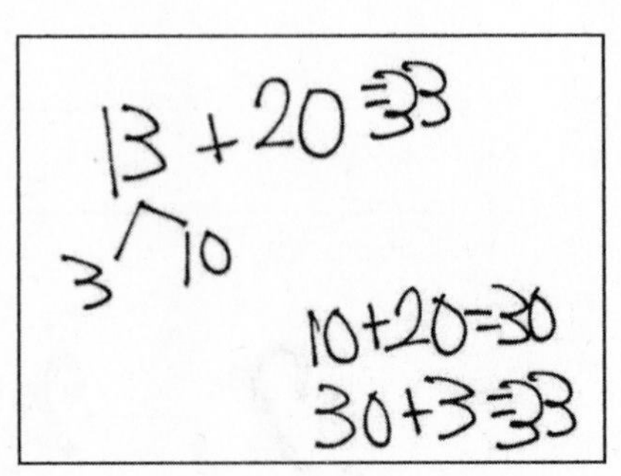

d. Fix the work that was incorrect by making a new drawing in the space below with the matching number sentence.

4. Solve using quick tens, the arrow way, or number bonds.

$$17 + 5 = \_\_\_\_$$

Share with your partner. Discuss why you chose to solve the way you did.

Name ____________________________ Date ______________

1. Two students both solved the addition problem below using different methods.

18 + 9

18 + 9 = 27
2 7
18 + 2 = 20
20 + 7 = 27

18 + 9 = 27
18 $\xrightarrow{+2}$ 20 $\xrightarrow{+7}$ 27
18 + 2 = 20
20 + 7 = 27

Are they both correct? Why or why not?

2. Another two students solved the same problem using quick tens.

18 + 9 = 29
20 + 9 = 29

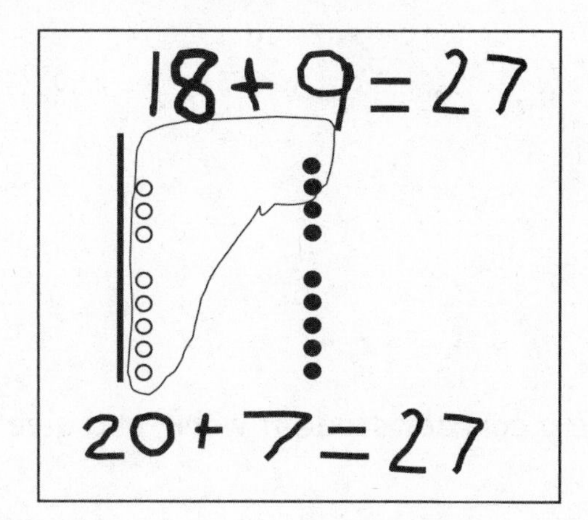

Are they both correct? Why or why not?

3. Circle any student work that is correct.

19 + 6

Student A	Student B	Student C
19 + 6 20 + 6 = 26	19 + 6 1 5 19 + 1 = 20 20 + 5 = 25	19 + 6 19 +1→ 20 +5→ 25

Fix the student work that was incorrect by making a new drawing or drawings in the space below.

Choose a correct student work, and give a suggestion for improvement.

Name ______________________________ Date ______________

Read the word problem.
Draw a tape diagram and label.
Write a number sentence and a statement that matches the story.

1. Lee saw 6 squashes and 7 pumpkins growing in his garden. How many vegetables did he see growing in his garden?

Lee saw __________ vegetables.

2. Kiana caught 6 lizards. Her brother caught 6 snakes. How many reptiles do they have altogether?

Kiana and her brother have __________ reptiles.

3. Anton's team has 12 soccer balls on the field and 3 soccer balls in the coach's bag. How many soccer balls does Anton's team have?

Anton's team has __________ soccer balls.

4. Emi had 13 friends over for dinner. 4 more friends came over for cake. How many friends came over to Emi's house?

There were __________ friends.

5. 6 adults and 12 children were swimming in the lake. How many people were swimming in the lake?

There were __________ people swimming in the lake.

6. Rose has a vase with 13 flowers. She puts 7 more flowers in the vase. How many flowers are in the vase?

There are __________ flowers in the vase.

Name ________________________________ Date ______________

Read the word problem.
Draw a tape diagram and label.
Write a number sentence and a statement that matches the story.

16
12 | 4

1. Darnel is playing with his 4 red robots. Ben joins him with 13 blue robots. How many robots do they have altogether?

They have ________ robots.

2. Rose and Emi had a jump rope contest. Rose jumped 14 times, and Emi jumped 6 times. How many times did Rose and Emi jump?

They jumped ________ times.

3. Pedro counted the airplanes taking off and landing at the airport. He saw 7 airplanes take off and 6 airplanes land. How many airplanes did he count altogether?

Pedro counted _______ airplanes.

4. Tamra and Willie scored all the points for their team in their basketball game. Tamra scored 13 points, and Willie scored 5 points. What was their team's score for the game?

The team's score was _______ points.

Name ______________________________ Date ______________

Read the word problem.
Draw a tape diagram and label.
Write a number sentence and a statement that matches the story.

16
12 | 4

1. 9 dogs were playing at the park. Some more dogs came to the park. Then, there were 11 dogs. How many more dogs came to the park?

__________ more dogs came to the park.

2. 16 strawberries are in a basket for Peter and Julio. Peter eats 8 of them. How many are there for Julio to eat?

Julio has __________ strawberries to eat.

3. 13 children are on the roller coaster. 3 adults are on the roller coaster. How many people are on the roller coaster?

There are __________ people on the roller coaster.

4. 13 people are on the roller coaster now. 3 adults are on the roller coaster, and the rest are children. How many children are on the roller coaster?

There are __________ children on the roller coaster.

5. Ben has 6 baseball practices in the morning this month. If Ben also has 6 practices in the afternoon, how many baseball practices does Ben have?

Ben has __________ baseball practices.

6. Some yellow beads were on Tamra's bracelet. After she put 14 purple beads on the bracelet, there were 18 beads. How many yellow beads did Tamra's bracelet have at first?

Tamra's bracelet had __________ yellow beads.

Name ______________________________ Date ______________

Read the word problem.
Draw a tape diagram and label.
Write a number sentence and a statement that matches the story.

1. Rose has 12 soccer practices this month. 6 practices are in the afternoon, but the rest are in the morning. How many practices will be in the morning?

Rose has ______ practices in the morning.

2. Ben caught 16 fish. He put some back in the lake. He brought home 7 fish. How many fish did he put back in the lake?

Ben put ______ fish back in the lake.

3. Nikil solved 9 problems on the first Sprint. He solved 11 problems on the second Sprint. How many problems did he solve on the two Sprints?

Nikil solved ______ problems on the Sprints.

4. Shanika returned some books to the library. She had 16 books at first, and she still has 13 books left. How many books did she return to the library?

Shanika returned ______ books to the library.

Name ______________________________ Date ______________

Read the word problem.
Draw a tape diagram and label.
Write a number sentence and a statement that matches the story.

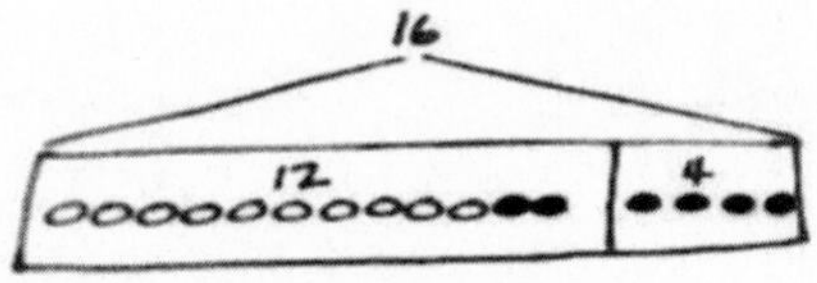

1. Rose drew 7 pictures, and Willie drew 11 pictures. How many pictures did they draw all together?

They drew ________ pictures.

2. Darnel walked 7 minutes to Lee's house. Then, he walked to the park. Darnel walked for a total of 18 minutes. How many minutes did it take Darnel to get to the park?

It took Darnel ________ minutes to get to the park.

3. Emi has some goldfish. Tamra has 14 betta fish. Tamra and Emi have 19 fish in all. How many goldfish does Emi have?

Emi has ________ goldfish.

4. Shanika built a block tower using 14 blocks. Then, she added 4 more blocks to the tower. How many blocks are there in the tower now?

The tower is made of __________ blocks.

5. Nikil's tower is 15 blocks tall. He added some more blocks to his tower. His tower is 18 blocks tall now. How many blocks did Nikil add?

Nikil added __________ blocks.

6. Ben and Peter caught 17 tadpoles. They gave some to Anton. They have 4 tadpoles left. How many tadpoles did they give to Anton?

They gave Anton __________ tadpoles.

Name ___________________________ Date ______________

Read the word problem.
Draw a tape diagram and label.
Write a number sentence and a statement that matches the story.

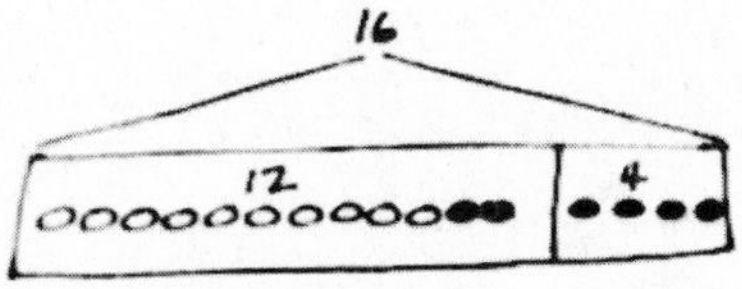

1. Fatima has 12 colored pencils in her bag. She has 6 regular pencils, too. How many pencils does Fatima have?

Fatima has ________ pencils.

2. Julio swam 7 laps in the morning. In the afternoon, he swam some more laps. He swam a total of 14 laps. How many laps did he swim in the afternoon?

Julio swam ______ laps in the afternoon.

3. Peter built 18 models. He built 13 airplanes and some cars. How many car models did he build?

Peter built ________ car models.

4. Kiana found some shells at the beach. She gave 8 shells to her brother. Now, she has 9 shells left. How many shells did Kiana find at the beach?

Kiana found _______ shells.

Name ________________________________ Date ________________

Use the tape diagrams to write a variety of word problems. Use the word bank if needed. Remember to label your model after you write the story.

Topics (Nouns)		
flowers	goldfish	lizards
stickers	rockets	cars
frogs	crackers	marbles

Actions (Verbs)		
hide	eat	go away
give	draw	get
collect	build	play

1.

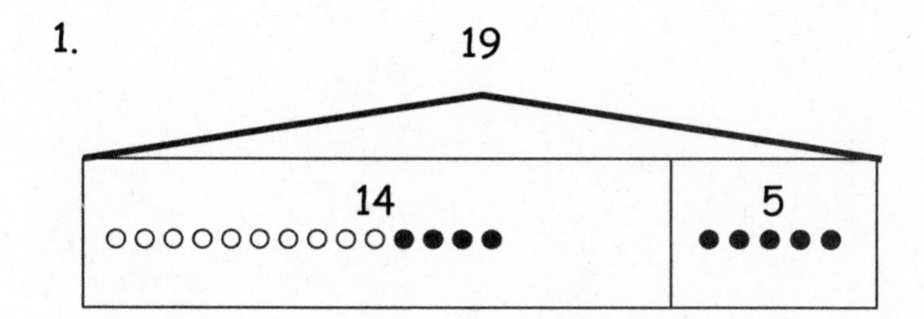

2.

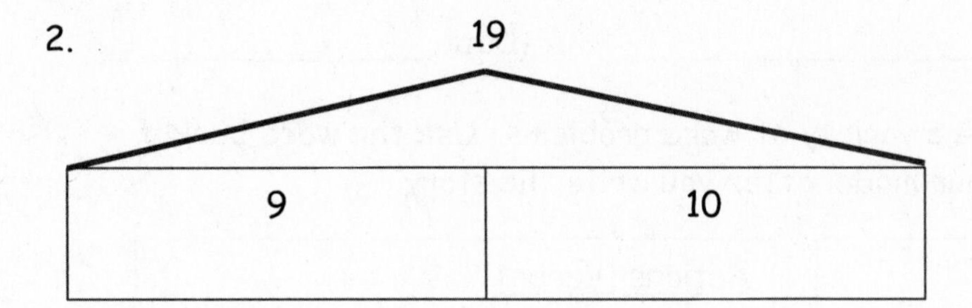

3.

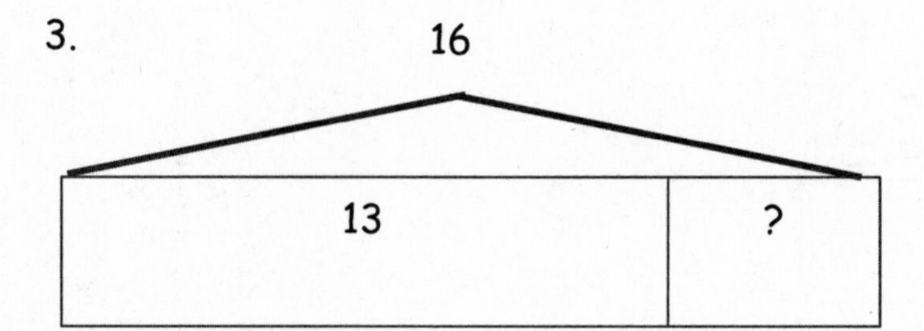

4.

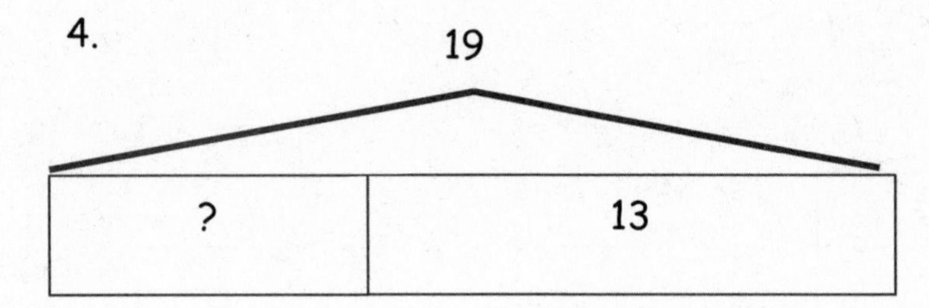

Name ______________________________ Date ______________

Use the tape diagrams to write a variety of word problems. Use the word bank if needed. Remember to label your model after you write the story.

Topics (Nouns)		
flowers	goldfish	lizards
stickers	rockets	cars
frogs	crackers	marbles

Actions (Verbs)		
hide	eat	go away
give	draw	get
collect	build	play

1.

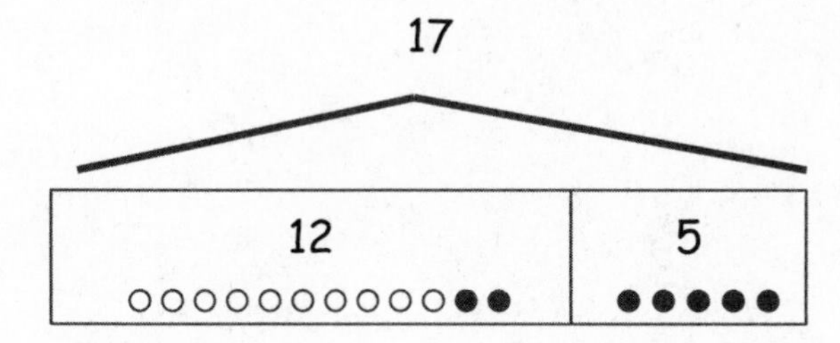

2.

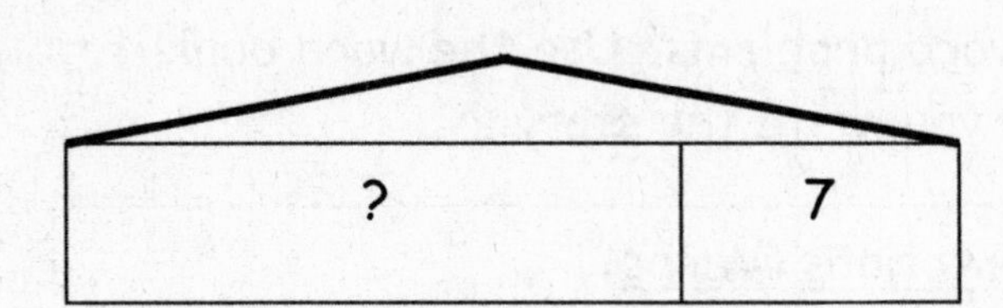

Name ______________________________ Date ______________

1. Fill in the blanks, and match the pairs that show the same amount.

a.

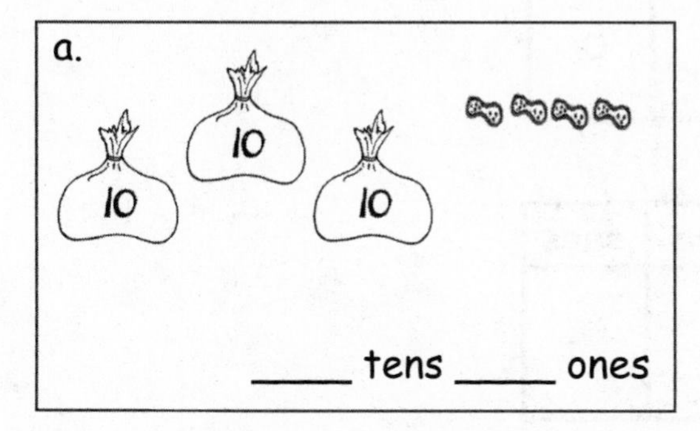

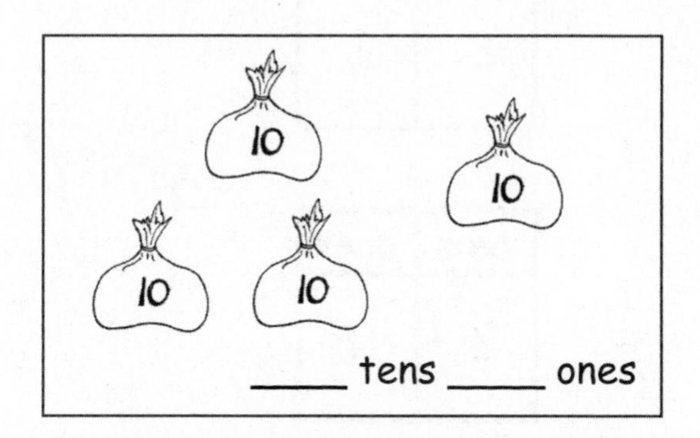

b.

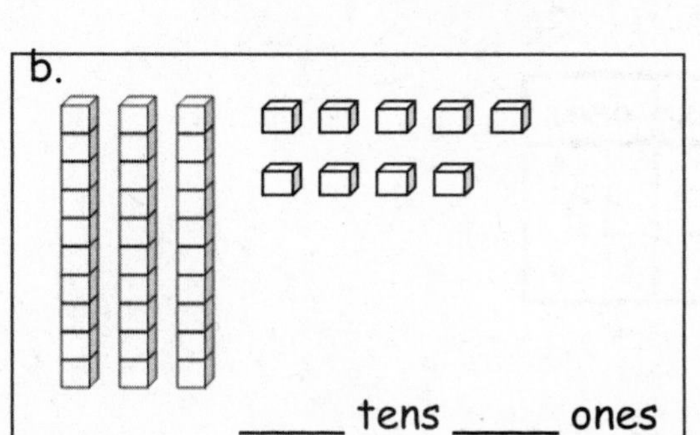

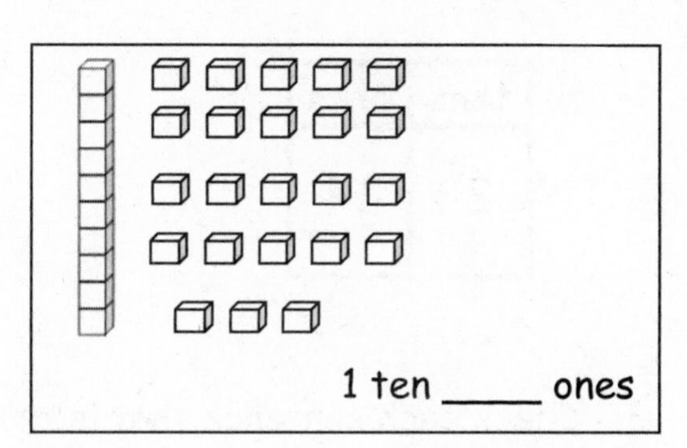

c.

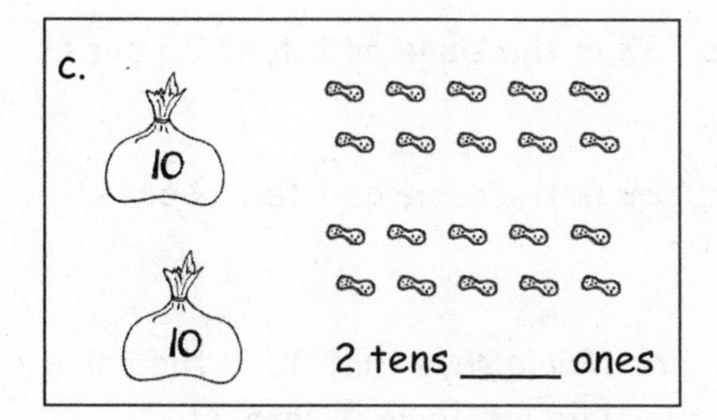

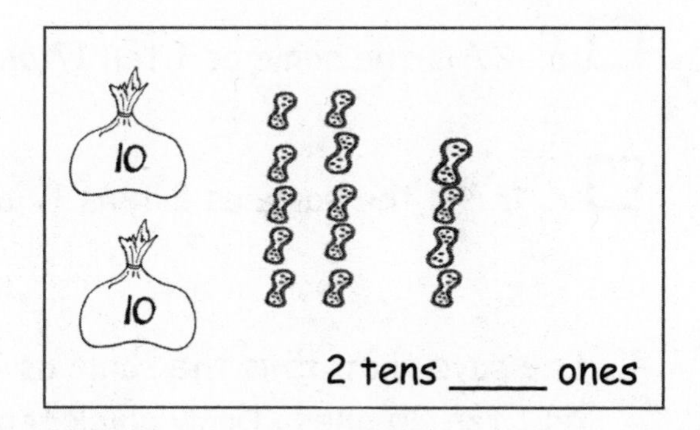

d.

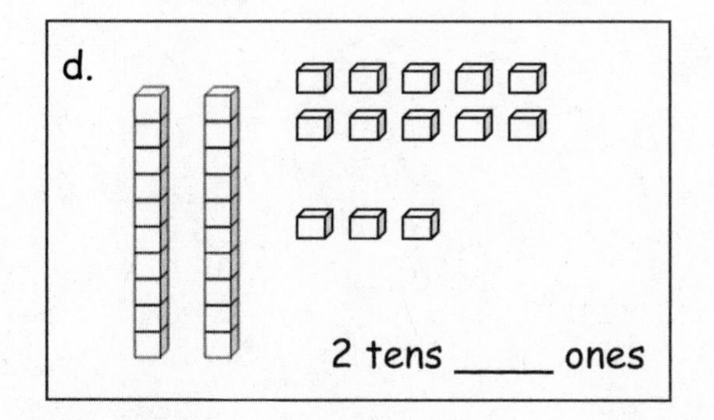

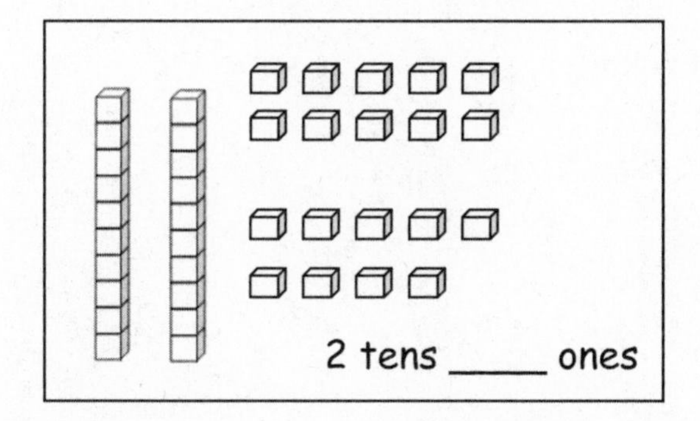

2. Match the place value charts that show the same amount.

a.

tens	ones
2	2

tens	ones
3	6

b.

tens	ones
2	16

tens	ones
3	4

c.

tens	ones
2	14

tens	ones
1	12

3. Check each sentence that is true.

☐ a. 27 is the same as 1 ten 17 ones.

☐ b. 33 is the same as 2 tens 23 ones.

☐ c. 37 is the same as 2 tens 17 ones.

☐ d. 29 is the same as 1 ten 19 ones.

4. Lee says that 35 is the same as 2 tens 15 ones, and Maria says that 35 is the same as 1 ten 25 ones. Draw quick tens to show if either Lee or Maria is correct.

Name __________________________________ Date ________________

1. Fill in the blanks, and match the pairs that show the same amount.

a.

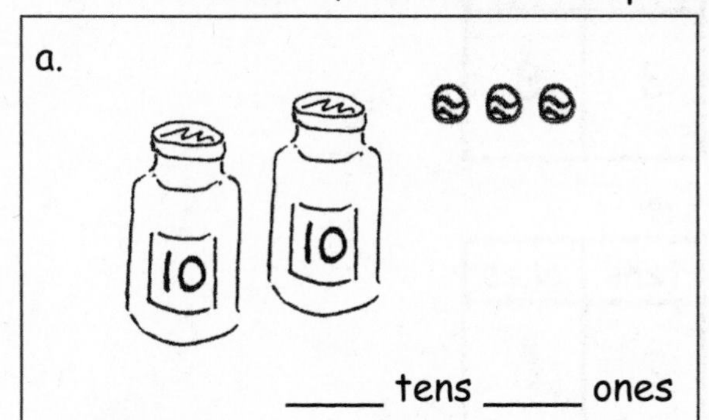

_____ tens _____ ones

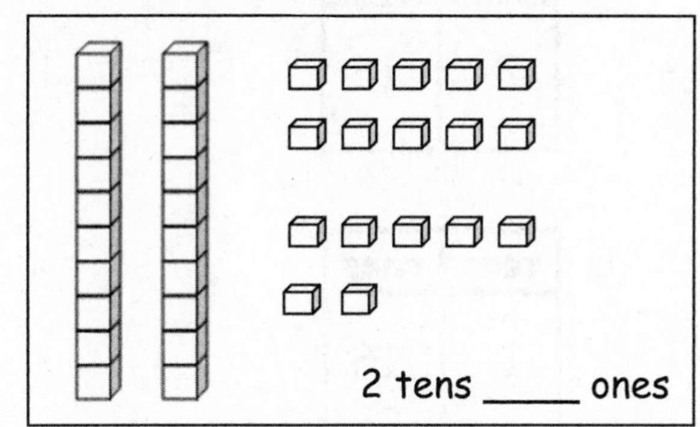

2 tens _____ ones

b.

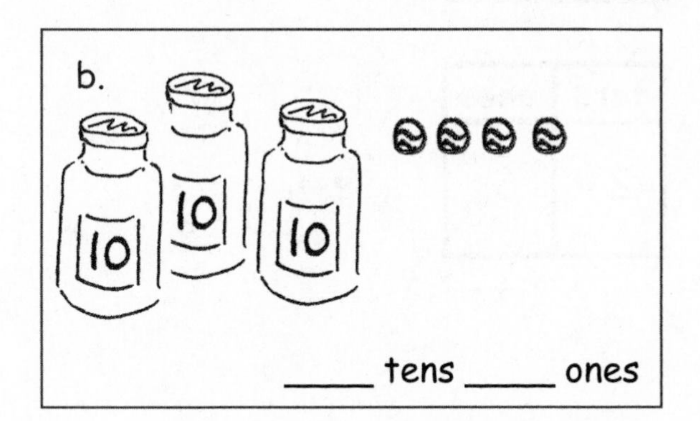

_____ tens _____ ones

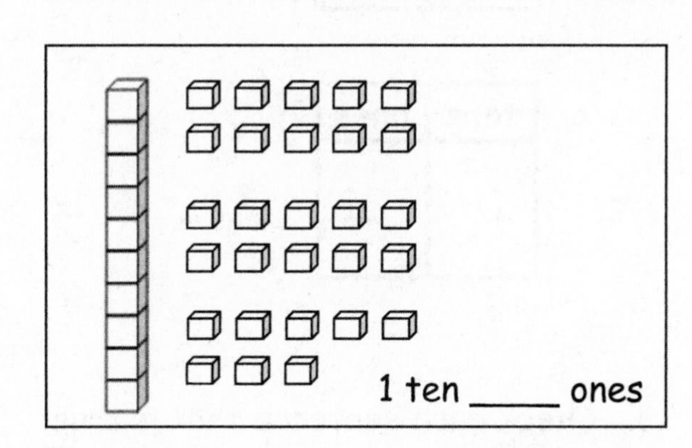

1 ten _____ ones

c.

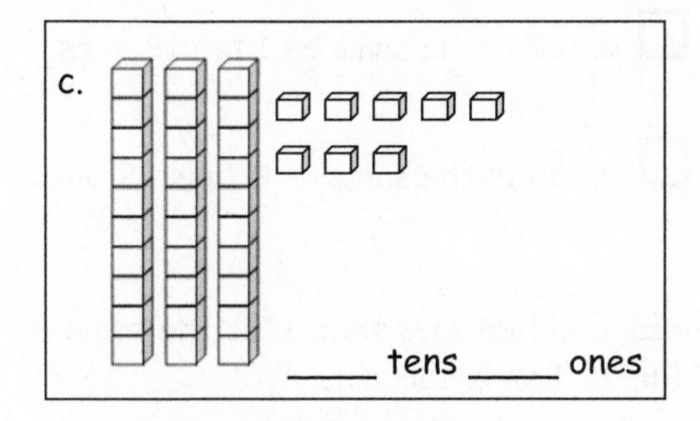

_____ tens _____ ones

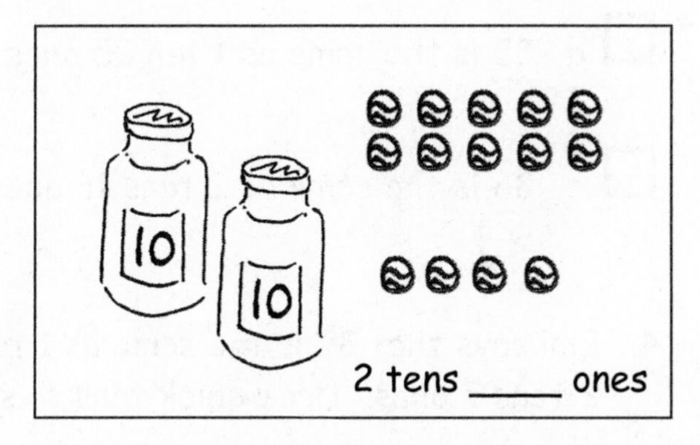

2 tens _____ ones

d.

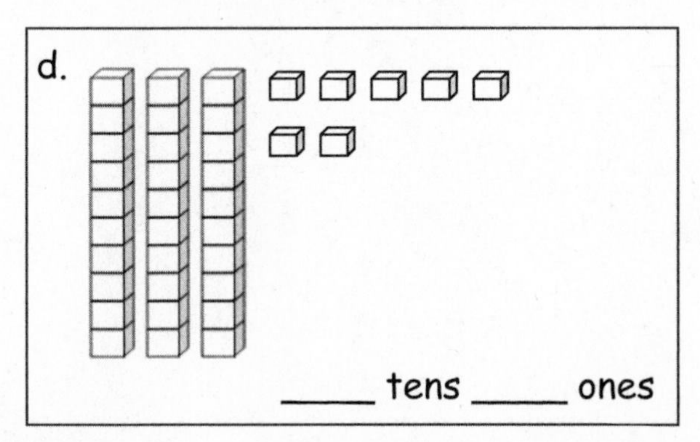

_____ tens _____ ones

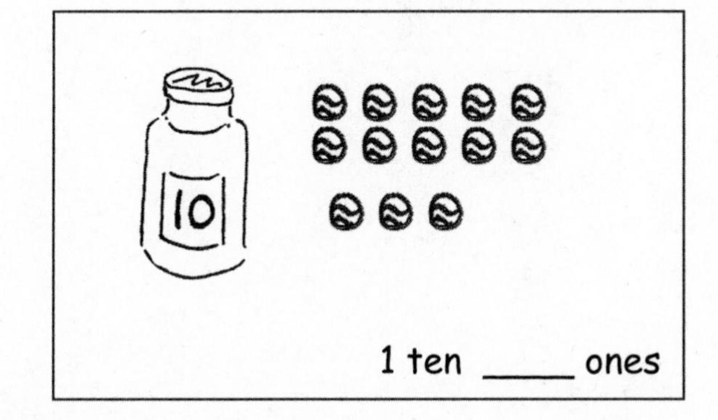

1 ten _____ ones

2. Match the place value charts that show the same amount.

a.

tens	ones
2	18

tens	ones
3	8

b.

tens	ones
1	16

tens	ones
2	1

c.

tens	ones
0	21

tens	ones
2	6

3. Check each sentence that is true.

☐ a. 35 is the same as 1 ten 25 ones.

☐ b. 28 is the same as 1 ten 18 ones.

☐ c. 36 is the same as 2 tens 16 ones.

☐ d. 39 is the same as 2 tens 29 ones.

4. Emi says that 37 is the same as 1 ten 27 ones, and Ben says that 37 is the same as 2 tens 7 ones. Draw quick tens to show if Emi or Ben is correct.

Name ______________________ Date ____________

1. Solve using number bonds. Write the two number sentences that show that you added the ten first. Draw quick tens and ones if that helps you.

a. 14 + 13 = ____ 10 3 14 + 10 = 24 24 + 3 = 27	b. 13 + 24 = ____ 10 3 24 + 10 = ____ ____ + 3 = ____
c. 16 + 13 = ____ 10 3 16 + 10 = ____ ____ + 3 = ____	d. 13 + 26 = ____ 10 3 26 + 10 = ____ ____ + ____ = ____
e. 15 + 15 = ____ 10 5 ____ + ____ = ____ ____ + ____ = ____	f. 15 + 25 = ____ ____ + ____ = ____ ____ + ____ = ____

2. Solve using number bonds or the arrow way. Part (a) has been started for you.

a. $15 + 13 =$ ____ 10 3	b. $14 + 23 =$ ____
c. $16 + 14 =$ ____	d. $14 + 26 =$ ____
e. $21 + 17 =$ ____	f. $17 + 23 =$ ____
g. $21 + 18 =$ ____	h. $18 + 12 =$ ____

Name ______________________________ Date ____________

1. Solve using number bonds. Write the two number sentences that show that you added the ten first. Draw quick tens and ones if that helps you.

a.

13 + 16 = ____

10 3

16 + 10 = 26

26 + 3 = 29

b.

16 + 23 = ____

10 6

23 + 10 = _____

_____ + 6 = _____

c.

16 + 14 = ____

10 4

16 + 10 = ____

____ + 4 = ____

d.

14 + 26 = ____

10 4

26 + 10 = ____

____ + ____ = ____

e.

17 + 13 = ____

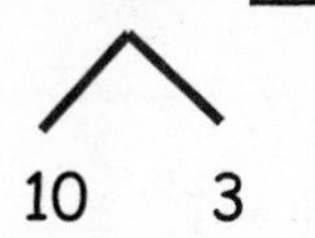

____ + ____ = ____

____ + ____ = ____

f.

27 + 13 = ____

____ + ____ = ____

____ + ____ = ____

2. Solve using number bonds. Part (a) has been started for you.

a. $14 + 13 =$ ____ 10 3 ____ + ____ = ____ ____ + ____ = ____	b. $24 + 14 =$ ____ ____ + ____ = ____ ____ + ____ = ____
c. $15 + 14 =$ ____	d. $24 + 15 =$ ____
e. $22 + 17 =$ ____	f. $27 + 12 =$ ____
g. $18 + 12 =$ ____	h. $28 + 12 =$ ____

Name ______________________________ Date ______________

1. Solve using number bonds. This time, add the tens first. Write the 2 number sentences to show what you did.

a. 11 + 14 = _____	b. 21 + 14 = _____
c. 14 + 15 = _____	d. 26 + 14 = _____
e. 26 + 13 = _____	f. 13 + 24 = _____

2. Solve using number bonds. This time, add the ones first. Write the 2 number sentences to show what you did.

a. 29 + 11 = _____	b. 17 + 13 = _____
c. 14 + 16 = _____	d. 26 + 13 = _____
e. 28 + 11 = _____	f. 12 + 27 = _____
g. 18 + 12 = _____	h. 22 + 18 = _____

Name ______________________ Date ____________

1. Solve using number bonds. This time, add the tens first. Write the 2 number sentences to show what you did.

a. $12 + 14 =$ ____	b. $14 + 21 =$ ____
c. $15 + 14 =$ ____	d. $25 + 14 =$ ____
e. $23 + 16 =$ ____	f. $16 + 24 =$ ____

2. Solve using number bonds. This time, add the ones first. Write the 2 number sentences to show what you did.

a. 27 + 10 = _____	b. 27 + 13 = _____
c. 13 + 26 = _____	d. 26 + 14 = _____
e. 12 + 18 = _____	f. 18 + 21 = _____
g. 19 + 11 = _____	h. 21 + 19 = _____

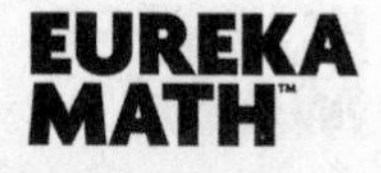

Name ______________________ Date __________

1. Solve using a number bond to add ten first. Write the 2 addition sentences that helped you.

a. $18 + 14 =$ ____ (14 → 10, 4) $18 + 10 = 28$ $28 + 4 = 32$	b. $14 + 17 =$ ____ (14 → 10, 4) $17 + 10 = 27$ $27 + 4 = 31$
c. $19 + 15 =$ ____ (15 → 10, 5) $19 + 10 =$ ____ ____ $+ 5 =$ ____	d. $18 + 15 =$ ____ (15 → 10, 5) $18 + 10 =$ ____ ____ $+ 5 =$ ____
e. $19 + 13 =$ ____ (13 → 10, 3) $19 + 10 =$ ____ ____ + ____ = ____	f. $19 + 16 =$ ____ (16 → 10, 6) $19 + 10 =$ ____ ____ + ____ = ____

2. Solve using a number bond to make a ten first. Write the 2 number sentences that helped you.

a.	b.
19 + 14 = ______ 1 13 19 + 1 = 20 20 + 13 = 33	18 + 13 = ______ 2 11 18 + 2 = 20 20 + 11 = 31
c. 18 + 14 = ______ 2 12 18 + 2 = _____ 20 + 12 = _____	d. 18 + 16 = _____ 2 14 18 + 2 = _____ _____ + 14 = _____
e. 15 + 17 = _____ 12 3 _____ + 3 = _____ _____ + 12 = _____	f. 17 + 18 = _____ 15 2 _____ + _____ = _____ _____ + _____ = _____

Name  Date ________

1. Solve using a number bond to add ten first. Write the 2 addition sentences that helped you.

a. $18 + 13 =$ ____ (13 → 10, 3) $18 + 10 = 28$ $28 + 3 = 31$	b. $13 + 19 =$ ____ (13 → 10, 3) $19 + 10 = 29$ $29 + 3 = 32$
c. $17 + 15 =$ ____ (15 → 10, 5) $17 + 10 =$ ____ ____ $+ 5 =$ ____	d. $17 + 16 =$ ____ (16 → 10, 6) $17 + 10 =$ ____ ____ $+ 6 =$ ____
e. $17 + 14 =$ ____ (14 → 10, 4) $17 + 10 =$ ____ ____ + ____ = ____	f. $19 + 17 =$ ____ (17 → 10, 7) $19 + 10 =$ ____ ____ + ____ = ____

2. Solve using a number bond to make a ten first. Write the 2 number sentences that helped you.

a. $19 + 13 = $ ____ (number bond: 1, 12) $19 + 1 = 20$ $20 + 12 = 32$	b. $19 + 14 = $ ____ (number bond: 1, 13) $19 + 1 = 20$ $20 + 13 = 33$
c. $18 + 15 = $ ____ (number bond: 2, 13) $18 + 2 = $ ____ $20 + 13 = $ ____	d. $18 + 17 = $ ____ (number bond: 2, 15) $18 + 2 = $ ____ ____ $+ 15 = $ ____
e. $18 + 19 = $ ____ (number bond: 17, 1) ____ $+ 1 = $ ____ ____ $+ 17 = $ ____	f. $19 + 19 = $ ____ (number bond: 18, 1) ____ + ____ = ____ ____ + ____ = ____

Name ______________________________ Date ______________

1. Solve using number bonds with pairs of number sentences. You may draw quick tens and some ones to help you.

a. 19 + 12 = _____	b. 18 + 12 = _____
c. 19 + 13 = _____	d. 18 + 14 = _____
e. 17 + 14 = _____	f. 17 + 17 = _____
g. 18 + 17 = _____	h. 18 + 19 = _____

2. Solve. You may draw quick tens and some ones to help you.

a. 19 + 12 = _____	b. 18 + 13 = _____
c. 19 + 13 = _____	d. 18 + 15 = _____
e. 19 + 16 = _____	f. 15 + 17 = _____
g. 19 + 19 = _____	h. 18 + 18 = _____

Name ____________________ Date __________

1. Solve using number bonds with pairs of number sentences. You may draw quick tens and some ones to help you.

a. 17 + 14 = ____	b. 16 + 15 = ____
c. 17 + 15 = ____	d. 18 + 13 = ____
e. 18 + 15 = ____	f. 18 + 16 = ____
g. 19 + 15 = ____	h. 19 + 16 = ____

2. Solve. You may draw quick tens and some ones to help you.

a. 19 + 14 = _____	b. 19 + 17 = _____
c. 18 + 17 = _____	d. 16 + 16 = _____
e. 17 + 14 = _____	f. 15 + 16 = _____
g. 19 + 19 = _____	h. 18 + 18 = _____

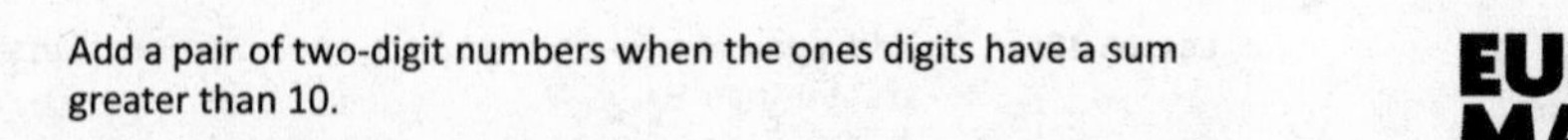

Name ______________________________ Date ______________

1. Solve using quick ten drawings, number bonds, or the arrow way. Check the rectangle if you made a new ten.

a. 23 + 12 = ____	b. 15 + 15 = ____
c. 19 + 21 = ____ 	d. 17 + 12 = ____
e. 27 + 13 = ____ 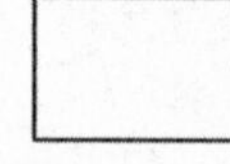	f. 17 + 16 = ____ 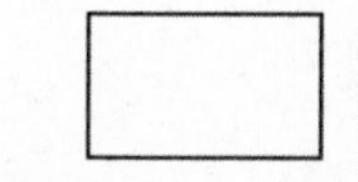

2. Solve using quick ten drawings, number bonds, or the arrow way.

a. 15 + 13 = _____	b. 25 + 13 = _____
c. 24 + 14 = _____	d. 25 + 15 = _____
e. 18 + 14 = _____	f. 18 + 18 = _____
g. 24 + 16 = _____	h. 17 + 18 = _____

Name ______________________________ Date ______________

Solve using quick tens and ones, number bonds, or the arrow way.

a. 13 + 16 = _____	b. 15 + 16 = _____
c. 16 + 16 = _____	d. 26 + 12 = _____
e. 22 + 17 = _____	f. 17 + 15 = _____
g. 17 + 16 = _____	h. 18 + 17 = _____

i. 24 + 13 = _____	j. 15 + 24 = _____
k. 19 + 16 = _____	l. 14 + 22 = _____
m. 27 + 12 = _____	n. 28 + 12 = _____
o. 18 + 17 = _____	p. 19 + 18 = _____

Name ______________________________ Date ______________

1. Solve using quick ten drawings, number bonds, or the arrow way.

a. 13 + 12 = ____	b. 23 + 12 = ____
c. 13 + 16 = ____	d. 23 + 16 = ____
e. 13 + 27 = ____	f. 17 + 16 = ____
g. 14 + 18 = ____	h. 18 + 17 = ____

2. Solve using quick ten drawings, number bonds, or the arrow way. Be prepared to discuss how you solved during the Debrief.

a. 17 + 11 = _____	b. 17 + 21 = _____
c. 27 + 13 = _____	d. 17 + 14 = _____
e. 13 + 26 = _____	f. 17 + 17 = _____
g. 18 + 15 = _____	h. 16 + 17 = _____

EUREKA MATH™